Lecture Notes in Control and Information Sciences   427

For further volumes:
http://www.springer.com/series/642

Yue Wang and Islam I. Hussein

Search and Classification Using Multiple Autonomous Vehicles

Decision-Making and Sensor Management

 Springer

Authors
Prof. Dr. Yue Wang
Department of Electrical Engineering
University of Notre Dame
Notre Dame
USA

Prof. Islam I. Hussein
Department of Mechanical Engineering
Worcester Polytechnic Institute
Worcester
USA

ISSN 0170-8643
ISBN 978-1-4471-2956-1
DOI 10.1007/978-1-4471-2957-8
Springer London Heidelberg New York Dordrecht

e-ISSN 1610-7411
e-ISBN 978-1-4471-2957-8

Library of Congress Control Number: 2012934375

Printed on acid-free paper

Springer is part of Springer Science+Business Media (www.springer.com)

This book is dedicated to my parents.

-Yue Wang

Preface

This book focuses on the real-time decision-making for large-scale domain search and object classification using Multiple Autonomous Vehicles (MAV). In recent years, MAV systems have attracted considerable attention and have been widely utilized. Of particular interest is their application to search and classification under limited sensory capabilities. Since search requires sensor mobility and classification requires a sensor to stay within the vicinity of an object, search and classification are two competing tasks. Therefore, there is a need to develop real-time sensor allocation decision-making strategies to guarantee task accomplishment. These decisions are especially crucial when the domain is much larger than the field-of-view of a sensor, or when the number of objects to be found and classified is much larger than that of available sensors.

In this book, the search problem is formulated as a coverage control problem, which aims at collecting enough data at every point within the domain to construct an awareness map. The object classification problem seeks to satisfactorily categorize the properties of each found object of interest. The decision-making strategies include both sensor allocation decisions and vehicle motion control. The awareness-, Bayesian-, and risk-based decision-making strategies are developed in sequence. The awareness-based approach is developed under a deterministic framework, while the latter two are developed under a probabilistic framework where uncertainty in sensor measurement is taken into account. The risk-based decision-making strategy also analyzes the effect of measurement cost. It is further extended to an integrated detection and estimation problem with applications in optimal sensor management. Simulation-based studies are performed to confirm the effectiveness of the proposed algorithms.

Chapter 1 provides an overview of the literature on MAV systems and their applications in domain search and object classification. The problem of decision-making and sensor management using MAVs is motivated to solve competing search and classification tasks under limited sensory resources. We provide a detailed study on coverage control in Chapter 2 under both deterministic and probabilistic frameworks. Corresponding to these coverage control formulations, we propose an awareness-based decision-making strategy in Chapter 3, a Bayesian-based decision-making strategy in Chapter 4, and a risk-based decision-making strategy in Chapter 5. The awareness-based strategy is constructed based on the real-time information about the competing search and classification tasks. It models the level of awareness of an autonomous vehicle about the events occurring within the mission domain. The Bayesian-based strategy extends the deterministic work in Chapter 3 by including uncertainties in sensor perception. Bayes filter and probability theory is utilized in this framework. Build upon this, the risk-based strategy further considers the cost of taking more observations for better decision-making using Bayesian sequential detection method. Its application on the space situational awareness problem is investigated in detail. In Chapter 6, we integrate Bayesian sequential detection and Bayesian sequential estimation method for the sensor management in domain search and object classification. A summary of future research directions is presented in the concluding chapter.

We would like to express our appreciation to Dr. Richard S. Erwin (Air Force Research Laboratory), Professor Donald R. Brown, Professor Michael A. Demetriou, Professor Stephen S. Nestinger (Worcester Polytechnic Institute), and Professor Yangquan Chen (Utah State University) for their valuable suggestions regarding the manuscript. We would also like to thank Springer Verlag and its staff for the professional support. Especially, we are grateful to Editor Oliver Jackson for his help and guidance in the development of this book.

Yue Wang   Islam I. Hussein
Electrical Engineering Department   *Mechanical Engineering Department*
University of Notre Dame   *Worcester Polytechnic Institute*

December 2011

Contents

List of Figures

List of Tables

Chapter 1
Introduction

1.1 Motivation and Objectives

In many domain search and object classification problems, including aerial search and rescue/destroy, surveillance, space imaging systems, mine countermeasures, and wildfire control, the effective management of limited available sensing resources is key to mission success [53, 144].

There are two basic objectives in a search and classification problem. The objective for domain search is to find each object of interest in a given domain and fix its position in space (and time for dynamic objects). The objective for object classification is to observe each found object until the desired amount of information has been collected to determine the property of the object. The characteristics of interest may include geometric shape, categorization, nature of electromagnetic emissions and object property. When the object is mobile, the objective is to track its state (e.g., position and velocity).

Given limited sensory capabilities, it is crucial to allocate resources (which vehicle should search/classify what?) and assign tasks (should a vehicle search or classify?) using the most efficient way possible. A sensor vehicle can perform either the search mission or the classification mission, but not both at the same time (search requires mobility and classification requires neighboring the object). On one hand, with limited available observations in the presence of sensor errors, a sensor may give a false alarm of object presence while there is actually none, miss detection of a critical object, or report incorrect classifications. On the other hand, taking exhaustive observations at one particular location of interest may result in losing the opportunity to find and classify possibly more critical objects at other locations within the domain. Hence, a vehicle sensor has to decide on whether to continue searching more unknown objects and sacrifice the decision accuracy, or keep taking observations at the current location and ignore elsewhere in the domain. This is especially true when 1) the size of the mission domain is much larger as compared to the limited sensory range of the vehicles, and 2) the number of unknown objects to be detected and classified is greater than that of available MAVs. Here, a large-scale

Y. Wang and I.I. Hussein: Search and Classification Using MAV, LNCIS 427, pp. 1–9.
springerlink.com

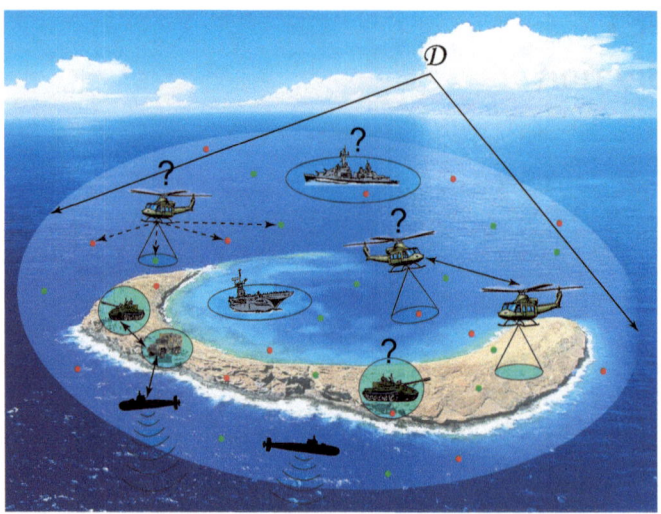

Fig. 1.1 Decision-making for search and classification.

domain is defined as a domain if a set of static limited-range sensors can not cover every point within it even in the worst case scenario when all the sensory ranges are disjoint.

This decision-making is critical in applications where one can not afford to search the whole space first and then classify, or classify until full certainty before proceeding with the search. For example, in search and rescue, if the vehicle sensor finds all potential human victims first, and then goes about classifying which are human victims and which are not, and decides to rescue only the classified human victims, by then, many victims could have passed away. On the contrary, if the vehicle sensor decides to classify each found object first with extremely high certainty before continuing to search, that may come at the cost of delaying the detection of critically injured victims who may pass away if not detected sooner. This also applies to scenarios where objects could be harmful (e.g., timed explosives) if their detection and classification is delayed. Therefore, there is a pressing need to develop MAV systems that seek to collect, process data and complete tasks efficiently under constrained resources.

Figure 1.1 illustrates a typical scenario for search and rescue using MAVs. Let $\mathscr{D}$ be a large-scale mission domain. The green and red dots represent unknown objects of interest to be found and classified. They are assumed to possess different properties and the number of objects is much larger than that of the available MAVs. Based on the progress of the search and rescue mission, each vehicle makes real-time decisions regarding whether to look for more objects within the domain, or keep taking observations at current locations. In this book, both deterministic and probabilistic decision-making strategies will be investigated to guarantee the detection and classification of all unknown objects of interest under such scenarios.

1.2 Literature Review

1.2.1 Multiple Autonomous Vehicle Systems

Many applications have emerged in recent years that rely on the use of a network of sensor-equipped MAVs to collect and process data [73, 66, 39, 114, 37]. This can be attributed to advances in relatively inexpensive and miniaturized networking and sensor technologies. The applications have widely spread over military, civilian and commercial areas and often involve tasks in adversarial and highly dynamic environments. In particular, MAVs have been increasingly used to perform operations that were traditionally carried out by humans, especially for missions that require operations in dangerous and highly dynamic environments that are hazardous to human operators. The advantages of autonomous vehicles over humans are (1) minimum risk of loss of human lives (e.g., search and rescue operations in hostile environments), and (2) more efficient computational power for data processing and real-time decision making as opposed to the limitations on human cognition, especially under stressful conditions. However, due to the limitations on computation and communication capabilities of a single on-board sensor, existing MAV systems are easily overwhelmed when dealing with large-scale information management. This then opens a niche for the current research on intelligent decision-making and task allocation scheme under limited sensory resources of the MAV systems.

There is rich literature on the control and applications of MAV systems. The coordination of MAVs has been a significant field of research with a broad range of applications in mobile robotics, intelligent highways, air traffic control, satellite clusters and so on. [95] provides a survey of recent research and future directions in cooperative control of MAV systems. Specific areas of interest include formation control, cooperative tasking, rendezvous, coverage, and consensus.

Graph theory[48] has been widely utilized on this topic. In [35], the authors use the Laplacian of a formation graph and present a Nyquist-like criterion for unmanned aerial vehicles (UAVs) formation stabilization. In [100], the authors study the MAV distributed formation control problem using potential functions obtained naturally from the structural constraints of a desired formation. The work in [65] focuses on the attitude alignment of MAVs using nearest neighbor rules. In [101], the multi-agent consensus problem is addressed under either fixed or switching topology, directed or undirected information flow in the absence or presence of communication delays.

Optimization-based approach is another large category of techniques for MAVs coordinated control. In [84], a decomposition team-optimal strategy is proposed for the rendezvous of multiple UAVs at a predetermined target location. The objective is to maximize the survivability of the UAVs. In [31], the MAV optimal formation control problem is investigated using receding horizon control. Mixed-integer linear programming (MILP) method has also been used for MAVs coordination problems because of its modeling capability and available commercial softwares [9, 32, 4].

The information-theoretic methods are well established, which seeks to maximize the information measures[49, 51].

Apart from the above work, the areas of particular interest in this book include using MAV systems for domain search and object classification, as well as the management of sensory resources. Section 1.2.2 discusses the work on coverage control, object detection, classification and tracking with MAVs. The literature on task assignment and sensor management using MAV systems will be provided in Section 6.1 of Chapter 6.

1.2.2   Domain Search, Object Classification and Tracking

The problem of domain search, unknown object classification and tracking has wide applications on humanitarian as well as military operations. Examples include but are not limited to the search-and-rescue operations in the open sea or sparsely populated areas [40], search and destroy missions for previously spotted enemy targets in multi-targeting/multi-platform battlefield [118], terrain acquisition [79], multi-agent (in particular, satellite) imaging systems [57], intelligent highway/vehicle systems [22], fire detection and prevention [26], mine clearing [27], room vacuuming [97], and lawn mowing [1].

Domain search deals with the problem of unknown object search and detection within a given domain. This problem usually requires the MAV systems to sense all reachable areas in the mission domain to achieve some objective function, e.g., minimum amount of time, maximum information, shortest path, etc. (see, for example [20, 2] and references therein). In [10], an excellent survey of the major results in search theory is provided. The problem of complete search for a target is studied in [55, 116, 130]. In [147], a probabilistic approach for domain search and path planning is proposed with multiple UAVs under global communications. The objective is to minimize the environment uncertainty in a finite amount of search time. The uncertainty map is updated using the Dempster-Shafer evidential method via sensor fusion. With the same goal, in [119], the authors present an agent-based negotiation scheme for a multi-UAV search operation with limited sensory and communication ranges. In [8], the problem of searching an area containing both regions of opportunity and hazard with multiple cooperative UAVs is considered. An alternate approach for searching in an uncertain environment is Simultaneous Localization and Mapping (SLAM) [74]. In [124], the occupancy grid mapping algorithm is addressed, which generates maps from noisy observations given foreknown robot pose. It is often used after solving a SLAM problem to generate robot navigation path from the raw sensor endpoints. In the robotics literature, a significant amount of research can be found in the field of robot motion planning [71, 79] and coverage path planning [120, 145, 146, 47, 68].

In this book, domain search is treated as the coverage control problem using sensor networks [15, 6, 134], where the vehicle sensor is controlled in such a way that the entire search domain can be covered. There are three major categories in coverage control, that is, optimal localization of immobile vehicle sensors, optimal redeployment/reconfigration of mobile sensors, and dynamic coverage control using mobile sensors. Under the scenario of large-scale mission domains where vehicle mobility is required, the third class of coverage control approach is adopted in this book. The goal is to dynamically cover every point within the domain using MAVs mounted with on-board sensors until achieving full coverage/awareness of the search environment. This problem is closely related to the coverage path planning problem in robotics. More details are provided in Chapter 2.

Unknown object classification and tracking together with domain search are generally treated as concurrent tasks that require the cooperation and/or decision-making of MAVs. Section 1.2.3 will provide a more detailed review of existing literature in this area and the comparison with the strategies proposed in this book. This section first reviews some related work with focus on classification and tracking. In [17], the authors address the problem of cooperative target classification using distributed UAVs. The views of neighboring vehicles are stochastically combined to maximize the probability of correct classification. In [16], the authors further discuss the capacitated transhipment and market-based bidding approaches to vehicle assignment for cooperative classification. A similar cooperative classification scheme is discussed in [64] for munition problems, which aids in reducing the false target attack rate. A binary object classification model is presented in [36], which a task load balancing scheme is proposed to cope with the uncertain results in task completion. In [42], a heuristic agreement strategy is presented for the cooperative control of UAVs. The authors associate a classification difficulty with each target, and a classification effort with each UAV. In [91], teams of UAVs are utilized to track moving targets in a cooperative convoy escort mission. The UAVs follow the convoy based on the shared GPS information from the ground vehicles and track suspicious attackers based on the live video. A hierarchical, hybrid control architecture is proposed for the cooperative tracking strategy. In [77], a class of collaborative signal processing techniques is investigated with focus on a vehicle tracking application using sensor networks. A leader-based information-driven tracking scheme is presented, which enables energy-efficient sensor selection. In [85], a cooperative acoustic tracking method is presented using binary-detection sensor networks. The tracking algorithm records the detection time of each sensor and performs line fitting for object's position estimates. The work in [69] discusses the trajectory tracking problem that requires the collective centroid of a group of nonholonomic UAVs to travel at a reference velocity. A cooperative tracking mechanism using multiple mobile sensors is provided in [33]. Detected targets are clustered using K-means clustering technique to minimize the number of required mobile sensors. An Extended Kohonen neural network is used as the tracking algorithm and an auction-based consensus mechanism is used as the cooperative strategy between trackers. In [92], a probabilistic tracking approach based on Condensation algorithm is proposed. Multiple pan-tilt-zoom cameras are used to track the objects with a level of reliability for belief updates.

1.2.3   Decision-Making for Search, Classification and Tracking

Although the literature in domain search, object classification and tracking is rich, little attention has been paid to the real-time decision-making for tasks competing for the same set of limited sensory resources.

Coordinated search and classification/tracking has been studied mainly for optimal path planning and state estimation in the literature. In [118], a distributed sequential auction scheme is presented for a multi-robot search and destroy operation. Local communications between neighbors are allowed and the shared information is used to make the decision. The control goal is to allocate an agent to an object and complete the mission in minimum time. Inspired by work on particle filter, in [117] the authors develop a strategy to dynamically control the relative configuration of sensor teams in order to get optimal estimates for target tracking through multi-sensor fusion. In [12, 13], the authors use the Beta distribution to model the level of confidence of target existence for an UAV search task. The minimum number of observations needed to achieve a probability threshold is derived. In [114], a cooperative control scheme based on Fischer information measure is proposed for the optimal path planning of a team of UAVs in a ground target tracking problem. In [127], a pursuit-evasion game and map building are combined in a probabilistic game theoretic framework, where sub-optimal pursuit policies are presented to minimize the expected capture time. In [81], the author proposes a Bayesian-based multitarget-multisensor management scheme. The approximation strategy, based on probability hypothesis densities, seeks to maximize the the expected number of targets. In [39], the target existence probability gain from searching a point is used as a cost function to determine the vehicle's optimal path. In [104], the control goal is to maximize the total number of observed objects and the amount of observation time of each. In [73], an optimal decision policy for the routing and munitions management of multiple formations of unmanned combat vehicles is proposed with imperfect information about adversarial ground units. A Recursive Bayesian Filter (RBF) is used for environment estimate. The threat type and location probabilities of the ground units are taken into account for classification. However, the underlying assumption made in the above research is that there is only a single object to be found, classified and tracked, or, the search domain is small and the sensing resources are not limited and, thus are not a concern.

The development of a unified framework for search and tracking problems has also been studied in several papers [40, 41, 125]. In [40], the authors investigate search-and-tracking using grid-based RBF with foreknown multi-target positions, but in the presence of noise. The results are extended in [72] to dynamic search spaces based on forward reachable set analysis. In [41], an element-based method is provided for a continuous probability density function of the target. In [125], the authors employ both grid-based Bayes filter and a particle filter for better estimation precision.

However, none of the above work considers search and classification/tracking as competing tasks, i.e., the tasks are equally-prioritized and do not need to compete for sensory resources. Considering the practical constraints of MAVs, it is motivated to develop a real-time treatment of unknown object search and classification missions, dealing with them as tasks competing for limited sensory resources.

Some related work is presented in [66], which considers a search-and-response mission with both known and unknown targets. The effects of weighting on search and classification is studied. The tradeoff between search and predicting task assignment is shown to be affected by the vehicles' resources and knowledge of target locations. This book considers the issue of limited resources, but focuses on optimal task assignment and hence still considers search and response in a unified framework as opposed to competing tasks. In [110], a survey of various approximate algorithms for Partially Observable Markov Decision Processes (POMDP) is provided for sequential decision-making in stochastic domains. POMDP methods cope with both the uncertainty in control actions and sensor errors. The RockSample problem is introduced to test these algorithms, where a robot chooses one of the actions ("move", "sample" and "check") to maximize rewards. Similarly, in [96], the authors present an approach for resource allocation in a cooperative air vehicle swarm system. The task allocation among search, classification and attack is modeled as a network flow optimization problem, which aims to maximize some global value.

1.3 Book Outline

The book is organized as follows. Chapter 2 introduces the coverage control problem. A review of the literature in coverage control is provided. Dynamic coverage control and awareness coverage control approaches are developed under a deterministic framework. Subsequently, Bayesian-based coverage control approaches are developed under a probabilistic framework. Underwater optical and acoustic seafloor mapping applications are discussed. In Chapter 3, an awareness-based decision-making strategy is proposed for search and classification based on the awareness coverage control laws developed in Chapter 2. In Chapter 4, a Bayesian-based probabilistic decision-making strategy is developed to take into account sensor errors. To further incorporate the cost of taking new observations, in Chapter 5, a risk-based sequential decision-making strategy is presented via Bayesian sequential detection. The binary results are extended to a more general ternary setting and its application to Space Situational Awareness (SSA) is investigated. In Chapter 6, the Bayesian sequential detection method for discrete random variables is extended to the Bayesian sequential estimation method for continuous random variables. The integration of these two approaches provides an optimal sensor management scheme that results in minimum information risk. The book is concluded with a summary of current and future work in Chapter 7.

1.4   Research Contributions

The major contribution and novelty of this book lies in the explicit treatment of search and classification as competing tasks based on the mission progress using MAVs with limited sensory ranges. The problems of domain search (coverage control), decision-making between search and classification (both deterministic and probabilistic), and the integration of detection and estimation with applications in SSA and underwater imaging are investigated.

The dynamic coverage control problem is first reviewed in Chapter 2. This lays a foundation for all the domain search methods in the decision-making strategies developed in this book.

Borrowing from the concept of dynamic coverage control, an awareness-based model is first proposed in Section 2.3 to describe how "aware" the vehicle sensors are of the environment. Both centralized and decentralized coverage control strategies are developed under global and intermittent communication architectures in Sections 2.3.4 and 2.3.3. Together with the classification strategy developed in Chapter 3, the awareness-based decision-making strategy guarantees the detection of all the unknown objects and the classification of each found object for at least a desired amount of time under limited sensory resources in a deterministic framework.

In order to take into account the uncertainty in sensor perception, in Section 2.4, a probabilistic coverage control strategy based on Bayes filter and information theory is developed. These results are extended in Section 2.4.5 to the case of MAVs with intermittent information sharing. A rigorous mathematical proof of the convergence of the expected probability of object existence is also provided. Coupling the search and classification processes, Chapter 4 proposes a Bayesian-based decision-making strategy that guarantees the detection and classification of all unknown objects in the presence of sensor errors.

Extending the Bayesian-based strategy, a risk-based decision-making strategy is proposed in Chapter 5 to take into account the cost of taking observations. The standard binary sequential detection method is utilized for risk analysis. Section 5.3 extends the result to a ternary setting which allows concurrent detection and classification observations. It is then applied to the SSA problem for the detection and classification of space objects in Earth orbit using a Space-Based Space Surveillance (SBSS) network in Section 5.4.

To further enable integrated detection and estimation decision-making, Chapter 6 proposes a risk-based sensor management scheme. The sequential estimation method is developed in Section 6.3 for the estimation of a process. Section 6.4 extends the risk analysis and decision making to the multi-element case based on both sequential detection and estimation methods. A risk-based optimal sensor management is then proposed in Section 6.5. The Rényi information measure is introduced to model the relative information loss in making a suboptimal sensor allocation decision, which is modeled as the observation cost in this book.

The following provides a list of publications that resulted from this book:

Journal Papers

1. Wang, Y., Hussein, I.I., Erwin, R.S.: Risk-based Sensor Management for Integrated Detection and Estimation. AIAA Journal of Guidance, Control, and Dynamics, JGCD (December 2010) (submitted)
2. Wang, Y., Hussein, I.I., Brown III, D.R., Erwin, R.S.: Cost-Aware Sequential Bayesian Decision-Making for Search and Classification. IEEE Transactions on Aerospace and Electronic Systems, TAES (2010) (under revision)
3. Wang, Y., Hussein, I.I.: Bayesian-Based Decision-Making for Object Search and Classification. IEEE Transactions on Control Systems Technology, TCST (2010) (in press)
4. Wang, Y., Hussein, I.I.: Awareness Coverage Control over Large-Scale Domains with Intermittent Communications. IEEE Transactions on Automatic Control (TAC) 55(8), 1850–1859 (2010)

Conference Papers

5. Wang, Y., Hussein, I.I., Erwin, R.S.: Sensor Management for Integrated Search and Tracking via Bayesian Sequential Analysis. In: American Control Conference (ACC), San Francisco, CA (June/July 2011)
6. Wang, Y., Hussein, I.I.: Multiple Vehicle Bayesian-Based Domain Search with Intermittent Information Sharing. In: American Control Conference (ACC), San Francisco, CA (June/July 2011)
7. Wang, Y., Hussein, I.I., Brown, D.R., Erwin, R.S.: Cost-Aware Bayesian Sequential Decision-Making for Domain Search and Object Classification. In: IEEE Conference on Decision and Control (CDC), Atlanta, GA, December 2010, pp. 7196–7201 (2010)
8. Wang, Y., Hussein, I.I., Erwin, R.S.: Bayesian Detection and Classification for Space-Augmented Space Situational Awareness under Intermittent Communications. In: Military Communications Conference (MILCOM), October 2010, pp. 960–965. San Jose, CA (2010) (Invited paper)
9. Wang, Y., Hussein, I.I., Brown III, D.R., Erwin, R.S.: Cost-Aware Sequential Bayesian Tasking and Decision-Making for Search and Classification. In: American Control Conference (ACC), Baltimore, MD, June/July 2010, pp. 6423–6428 (2010)
10. Wang, Y., Hussein, I.I.: Bayesian-Based Decision Making for Object Search and Characterization. In: American Control Conference (ACC), St. Louis, MO, June 2009, pp. 1964–1969 (2009)
11. Wang, Y., Hussein, I.I.: Underwater Acoustic Imaging using Autonomous Vehicles. In: IFAC Workshop on Navigation, Guidance and Control of Underwater Vehicles (NGCUV), Killaloe, Ireland (April 2008) (Invited Paper)
12. Wang, Y., Hussein, I.I., Erwin, R.S.: Awareness-Based Decision Making for Search and Tracking. In: American Control Conference (ACC), Seattle, WA, June 2008, pp. 3169–3175 (2008) (Invited Paper)
13. Wang, Y., Hussein, I.I.: Awareness Coverage Control over Large Scale Domains with Intermittent Communications. In: American Control Conference (ACC), Seattle, WA, June 2008, pp. 4370–4375 (2008)
14. Wang, Y., Hussein, I.I.: Vision-Based Coverage Control for Underwater Sampling using Multiple Submarines. In: IEEE Multiconference on Systems and Control (MSC) Covering IEEE CCA 2007 & IEEE ISIC 2007, Singapore, October 2007, pp. 82–87 (2007) (Invited Paper)

Chapter 2
Coverage Control

2.1 Cooperative Coverage Control

Coverage control studies the problem of covering a given search domain using
MAVs. In the literature of cooperative coverage control, a significant amount of
research can be found in two main categories: 1) Optimal localization of immobile
sensors [29, 99, 30], and 2) optimal redeployment of mobile sensors [78, 23, 76].

2.1.1 Location Optimization of Immobile Sensors

This class of problems only requires the distribution of a fixed sensor network in
the domain. The two variables of interest are sensor domains (the domain which
each sensor is responsible of sampling) and sensor locations. This algorithm can be
calculated off-line and no further mobility is required for the vehicles. The solution
is based on Voronoi partitions and the Lloyd algorithm [78]. The optimal sensor domain is a Voronoi cell in the partition and the optimal sensor location is its centroid
[30]. For a complete discussion of the coverage control problem applying Voronoi
partitions, see [24], where the authors propose both continuous and discrete-time
versions of the classic Lloyd algorithms for MAVs performing distributed sensing
tasks. In [86], a coverage control scheme based on Voronoi diagram is proposed to
maximize target exposure in some surveillance applications.

2.1.2 Optimal Redeployment of Mobile Sensors

The sensor redeployment problems involve the coordinated movement of MAVs for
an optimal final configuration. In [43], the authors provide a summary on current
control theories using MAV sensor networks. The coverage deployment problem

Y. Wang and I.I. Hussein: Search and Classification Using MAV, LNCIS 427, pp. 11–67.
springerlink.com   © Springer-Verlag London Limited 2012

aims at maximizing the area within a close range of mobile agents and uses a Voronoi partition algorithm. In [17], the authors use a Voronoi-based polygonal path approach and aim at minimizing exposure of a UAV fleet to radar. In [24], a dynamic version of the Lloyd algorithm is also provided. It drives each sensor to a unique centroid of a cell in a dynamic Voronoi partition of the search domain and iteratively achieves the optimal configuration. However, Voronoi-based approaches require exhaustive computational effort to compute the Voronoi cells continuously during a real-time implementation of the controllers. In [76], the authors develop an optimization problem that aims at maximizing coverage using sensors with limited ranges, while minimizing communication cost using a probabilistic network model. This class of problems is related to the active sensing literature in robotics [87], where Kalman filter is extensively used to process observations and generate estimates.

2.1.3 Dynamic Cooperative Coverage Control

An implicit assumption made in the above problem classes is that the mission domain is small-scale, i.e., one where in the best case scenario that can be covered by the union of a set of static limited-range sensors. This is equivalent to the assumption of infinite sensory ranges in the existing literature on the redeployment problem, which is especially true for work within the stochastic framework (see, for example, [49]) that assumes Gaussian distributions. However, this is not the case in many practical applications, where the field-of-view of the on-board sensors is relatively limited as compared to the size of the search domain, or there are too few sensor vehicles. For such problems, vehicle mobility is necessary to be able to account for all locations contained in the domain of interest and meet the coverage goal. Aside from large-scale domains, constantly moving sensors are also required for cases where sensors are mounted on mobile vehicles incapable of having zero velocities (e.g., fixed-wing aircraft), or when the host vehicles' safety is compromised if left fixed in space. Mobility of the vehicles in all these problems is also required since information of interest that is distributed over the domain may be changing in time. Not being able to continuously monitor parts of the domain for all time results in the requirement that the network is in a constant state of mobility with well-managed revisiting of locations in the domain to guarantee satisfactory awareness levels over the entire domain.

Dynamic cooperative coverage control is the vehicle motion control problem for coordinated MAVs to dynamically cover a given arbitrarily-shaped domain. The objective is to survey the entire search domain such that the information collected at each point achieves a preset desired amount. This is the fundamental difference between the two approaches presented above and dynamic coverage control, which is the method adopted for domain search in this book. While the aforementioned research focuses on the optimal or suboptimal configuration of MAVs to improve network coverage performance, dynamic cooperative coverage control guarantees that every point within the search domain will be sampled a desired amount of data with high certainty as a result of the constant movement of the MAVs.

Remark 2.1.1. *The key feature of the proposed approach is summarized as follows:*

- *The sensor is modeled to have a limited sensory range*
- *The dynamic coverage control strategy aims at collecting enough high quality data at each point in a domain of interest*

Applications include search and rescue missions where each point in the search domain has to be surveyed, aerial wildfire control in inaccessible and rugged country where each point in the wildfire region has to be "suppressed" using fixed-wing aircraft or helicopters, underwater sampling and mapping where each point in the deep ocean is required to be sufficiently sampled for marine geology, geophysics, biology, archaeology, and chemistry studies.

A slightly modified version of the coverage problem has been studied in [57] for (optimal and suboptimal) motion planning of multiple spacecraft interferometric imaging systems (MSIIS). The problem is also related to the literature on coverage path planning [20, 2] and Simultaneous Localization and Mapping (SLAM) [74] in robotics.

In the following sections, both deterministic and probabilistic vehicle motion control laws are developed for coverage control.

2.2 Deterministic Lyapunov-Based Approach

This section provides a brief summary of the major results of the coverage control problem discussed in [60]. It lays a foundation for all the search strategies presented in the subsequent sections. The vehicle collision avoidance and flocking control laws presented in [60] can also be applied to other search strategies discussed in this chapter via some straightforward modifications. Please refer to [60, 62] for more details.

2.2.1 Problem Formulation

Denote a vehicle by $\mathcal{V}$. Let $\mathbb{R}^+ = \{a \in \mathbb{R} : a \geq 0\}$, $Q = \mathbb{R}^2$ be the configuration space of all the vehicles and $\mathcal{D} \subseteq \mathbb{R}^2$ be the mission domain. Assume that $\mathcal{D}$ is a simply connected, bounded set with non-zero measure. Let the map $\phi : \mathcal{D} \to \mathbb{R}^+$, called a distribution density function, represent a measure of information or probability that some event takes place or object exists over $\mathcal{D}$. A large value of ϕ indicates high likelihood of event detection and a smaller value indicates low likelihood. Let N be the total number of MAVs and $\mathbf{q}_i \in Q$ denote the position of vehicle $\mathcal{V}_i$, $i \in \mathcal{S} = \{1, 2, 3, \ldots, N\}$. That is, the set $\mathcal{S}$ contains all vehicles performing the domain search task. Each vehicle $\mathcal{V}_i$, $i \in \mathcal{S}$, satisfies the following simple kinematic equations of motion

$$\dot{\mathbf{q}}_i = \mathbf{u}_i, \ i \in \mathcal{S}, \tag{2.1}$$

where $\mathbf{u}_i \in \mathbb{R}^2$ is the control velocity of vehicle $\mathcal{V}_i$. This is a simplified model and the results may be extended to agents with second order nonlinear dynamics evolving on more complex configuration manifolds.

Define the instantaneous coverage function $A_i : \mathcal{D} \times Q \to \mathbb{R}^+$ as a $\mathcal{C}^1$-continuous map that describes how effective a vehicle $\mathcal{V}_i$ senses a point $\tilde{\mathbf{q}} \in \mathcal{D}$. Let $s = \|\mathbf{q}_i(t) - \tilde{\mathbf{q}}\|$, which is the relative distance between the vehicle position and the measuring point. Without loss of generality, consider the following sensor model **SM**. This model is not an assumption for the ensuing theoretical results to be valid. The important feature of the proposed sensor model is that the sensors have a finite field-of-view.

Sensor Model SM.

1. Each vehicle has a peak sensing capacity of M_i exactly at the position $\mathbf{q}_i$ of vehicle $\mathcal{V}_i$, i.e., $s = \|\mathbf{q}_i(t) - \mathbf{q}_i(t)\| = 0$. That is,

$$A_i(0) = M_i > A_i(s), \ \forall s \neq 0.$$

2. Each vehicle sensor has a circular sensing symmetry about the position $\mathbf{q}_i$, $i \in \mathcal{S}$, in the sense that all points in $\mathcal{D}$ that are on the same circle centered at $\mathbf{q}_i$ are sensed with the same intensity. That is,

$$A_i(s) = \text{constant}, \ \forall s = c,$$

   for all constant c, $0 \leq c \leq r_i$, where r_i is the range of the sensor of vehicle $\mathcal{V}_i$.
3. Each vehicle has a limited *sensory domain*, $\mathcal{W}_i(t)$, with a *sensory range*, r_i. The sensory domain of each vehicle is given by

$$\mathcal{W}_i(t) = \{\tilde{\mathbf{q}} \in \mathcal{D} : \|\mathbf{q}_i(t) - \tilde{\mathbf{q}}\| \leq r_i\}. \tag{2.2}$$

Let the union of all coverage regions be denoted by

$$\mathcal{W}(t) = \cup_{i \in \mathcal{S}} \mathcal{W}_i(t).$$

An example of such a sensor function is a fourth order polynomial function of $s = \|\mathbf{q}_i(t) - \tilde{\mathbf{q}}\|$ within the sensor range and zero otherwise,

$$A_i(s) = \begin{cases} \frac{M_i}{r_i^4}\left(s^2 - r_i^2\right)^2 & \text{if } s \leq r_i \\ 0 & \text{if } s > r_i \end{cases}. \tag{2.3}$$

Figure 2.1 shows an instantaneous coverage function A_i (2.3) with $\mathbf{q}_i = (0,0)$, $M_i = 1$ and $r_i = 2$.

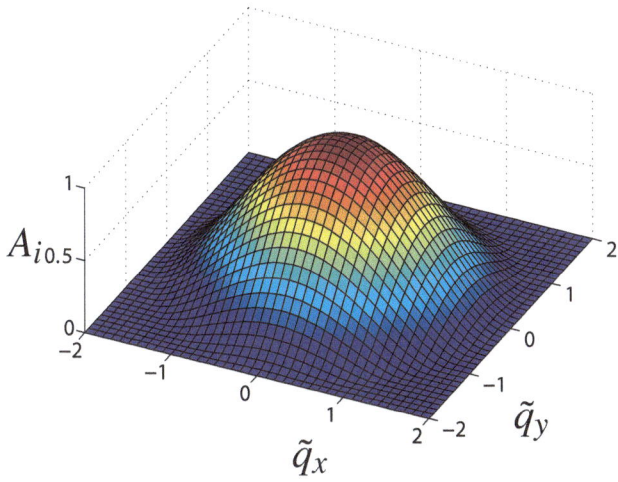

Fig. 2.1 Instantaneous coverage function.

Fixing a point $\tilde{\mathbf{q}}$, the effective coverage achieved by a vehicle $\mathscr{V}_i$ surveying $\tilde{\mathbf{q}}$ from the initial time $t_0 = 0$ to time t is defined to be

$$\mathscr{T}_i(\tilde{\mathbf{q}},t) := \int_0^t A_i(\|\mathbf{q}_i(\tau) - \tilde{\mathbf{q}}\|^2)\mathrm{d}\tau$$

and the effective coverage by a subset of vehicles $\mathscr{V}_{\mathscr{K}} = \{\mathscr{V}_j | j \in \mathscr{K} \subseteq \mathscr{S}\}$ in surveying $\tilde{\mathbf{q}}$ is then given by

$$\mathscr{T}_{\mathscr{K}}(\tilde{\mathbf{q}},t) := \sum_{i \in \mathscr{K}} \mathscr{T}_i(\tilde{\mathbf{q}},t) = \int_0^t \sum_{i \in \mathscr{K}} A_i(\|\mathbf{q}_i(\tau) - \tilde{\mathbf{q}}\|^2)\mathrm{d}\tau.$$

Note that $\mathscr{T}_{\mathscr{K}}(\tilde{\mathbf{q}},t)$ is a non-decreasing function of time t,

$$\frac{\partial}{\partial t}\mathscr{T}_{\mathscr{K}}(\tilde{\mathbf{q}},t) = \sum_{i \in \mathscr{K}} A_i\left(\|\mathbf{q}_i(t) - \tilde{\mathbf{q}}\|^2\right) \geq 0.$$

Let C^* be the desired attained effective coverage at all points $\tilde{\mathbf{q}} \in \mathscr{D}$. The goal is to attain an overall coverage of $\mathscr{T}_{\mathscr{S}}(\tilde{\mathbf{q}},t) = C^*$ for all $\tilde{\mathbf{q}} \in \mathscr{D}$ at some time t. The quantity C^* guarantees that, when $\mathscr{T}_{\mathscr{S}}(\tilde{\mathbf{q}},t) = C^*$, one can judge, with some level of confidence, whether or not an event occurs or an object exists at $\tilde{\mathbf{q}} \in \mathscr{D}$. Consider the following *error function*

$$e(t) = \int_{\mathscr{D}} h\left(C^* - \mathscr{T}_{\mathscr{S}}(\tilde{\mathbf{q}},t)\right)\phi(\tilde{\mathbf{q}})\mathrm{d}\tilde{\mathbf{q}}, \tag{2.4}$$

where $h(x)$ is a *penalty function* that satisfies $h(x) = h'(x) = h''(x) = 0$ for all $x \le 0$, and $h(x), h'(x), h''(x) > 0$ for all $x \in (0, C^*]$. The penalty function penalizes lack of coverage of points in $\mathcal{D}$. An example for the penalty function $h(x)$ is

$$h(x) = (\max(0, x))^2. \tag{2.5}$$

It incurs a penalty whenever $\mathcal{T}_{\mathscr{S}}(\tilde{\mathbf{q}}, t) < C^*$. Once $\mathcal{T}_{\mathscr{S}}(\tilde{\mathbf{q}}, t) \ge C^*$ at a point in $\mathcal{D}$, the error at this point is zero no matter how much additional time vehicles spend surveying that point. The total error is an average over the entire domain $\mathcal{D}$ weighted by the density function $\phi(\tilde{\mathbf{q}})$. When $e(t) = 0$, one says that the search mission is accomplished.

2.2.2 Vehicle Motion Control and Search Strategy

Without loss of generality, first consider the following initial condition whose utility will become obvious later:

IC1.   The initial coverage is identically zero: $\mathcal{T}_{\mathscr{S}}(\tilde{\mathbf{q}}, 0) = 0$, $\forall \tilde{\mathbf{q}} \in \mathcal{D}$.

Consider the following nominal control law

$$\bar{\mathbf{u}}_i(t) = -\bar{k}_i \int_{\mathcal{D}} h'(C^* - \mathcal{T}_{\mathscr{S}}(\tilde{\mathbf{q}}, t)) \left. \frac{\partial A_i(s^2)}{\partial(s^2)} \right|_{s = \|\mathbf{q}_i(t) - \tilde{\mathbf{q}}\|} \cdot (\mathbf{q}_i(t) - \tilde{\mathbf{q}}) \, \phi(\tilde{\mathbf{q}}) d\tilde{\mathbf{q}}, \tag{2.6}$$

where $\cdot$ denotes the inner product and $\bar{k}_i > 0$ are fixed feedback gains. Consider the function $\bar{V} = -e_t(t)$, where $e_t = \frac{de}{dt}$, and note that $\dot{V} = -e_{tt}$ where

$$e_t(t) = -\int_{\mathcal{D}} h'(C^* - \mathcal{T}_{\mathscr{S}}(\tilde{\mathbf{q}}, t)) \left(\sum_{j \in \mathscr{S}} A_j(\|\mathbf{q}_j(t) - \tilde{\mathbf{q}}\|^2) \right) \phi(\tilde{\mathbf{q}}) d\tilde{\mathbf{q}}$$

$$e_{tt} = \int_{\mathcal{D}} h''(C^* - \mathcal{T}_{\mathscr{S}}(\tilde{\mathbf{q}}, t)) \left(\sum_{j \in \mathscr{S}} A_j(\|\mathbf{q}_j(t) - \tilde{\mathbf{q}}\|^2) \right)^2 \psi(\tilde{\mathbf{q}}) d\tilde{\mathbf{q}}$$

$$-2 \int_{\mathcal{D}} h'(C^* - \mathcal{T}_{\mathscr{S}}(\tilde{\mathbf{q}}, t)) \left(\sum_{i \in \mathscr{S}} \left. \frac{\partial A_i(s^2)}{\partial(s^2)} \right|_{s = \|\mathbf{q}_i(t) - \tilde{\mathbf{q}}\|} (\mathbf{q}_j(t) - \tilde{\mathbf{q}}) \cdot \mathbf{u}_i \right) \phi(\tilde{\mathbf{q}}) d\tilde{\mathbf{q}}$$

$$= \int_{\mathcal{D}} h''(C^* - \mathcal{T}_{\mathscr{S}}(\tilde{\mathbf{q}}, t)) \left(\sum_{j \in \mathscr{S}} A_j(\|\mathbf{q}_j(t) - \tilde{\mathbf{q}}\|^2) \right)^2 \phi(\tilde{\mathbf{q}}) d\tilde{\mathbf{q}}$$

$$+2 \sum_{i \in \mathscr{S}} \bar{k}_i \left[\int_{\mathcal{D}} h'(C^* - \mathcal{T}_{\mathscr{S}}(\tilde{\mathbf{q}}, t)) \left(\left. \frac{\partial A_j(s^2)}{\partial(s^2)} \right|_{s = \|\mathbf{q}_j(t) - \tilde{\mathbf{q}}\|} \cdot (\mathbf{q}_j(t) - \tilde{\mathbf{q}}) \right) \phi(\tilde{\mathbf{q}}) d\tilde{\mathbf{q}} \right]^2$$

are the first and second time derivatives of $e(t)$ along the trajectory generated by the control law $\bar{\mathbf{u}}_i$ in Equation (2.6). Consider the following condition.

Condition C1. $\mathcal{T}_{\mathscr{S}}(\tilde{\mathbf{q}}, t) = C^*$, $\forall \tilde{\mathbf{q}} \in \mathscr{W}_i(t), \forall i \in \mathscr{S}$.

Lemma 2.2.1. *If for some $t \geq 0$ Condition **C1** holds, then $e_t(t) = 0$. Conversely, if $e_t(t) = 0$ for some time $t \geq 0$, then Condition **C1** holds.*

Proof. By the property **SM**3 of the sensor model, Condition **C1** implies that the h' term in the integrand in the expression for e_t is nonzero only outside $\mathscr{W}(t)$ where all coverage functions A_i are zero. That is, $h'\left(C^* - \mathscr{T}_{\mathscr{S}}(\tilde{\mathbf{q}},t)\right) = 0$ precisely inside $\mathscr{W}(t)$. Hence, under Condition **C1** $e_t = 0$.

The converse is easily verified by noting that the integrand in the expression for e_t is greater than or equal to zero everywhere in $\mathscr{D}$. For e_t to be zero, the integrand has to be identically equal to zero everywhere on $\mathscr{D}$, which holds true only if Condition **C1** holds. This completes the proof. □

From the lemma, $\bar{V} = -e_t \geq 0, \dot{\bar{V}} \leq 0$ with equality holding if and only if Condition **C1** holds. This implies that the function $\bar{V}$ is a Lyapunov-type function that guarantees that the system always converges to the state described in Condition **C1**. This proves the following result.

Lemma 2.2.2. *Under the control law (2.6), a MAV system will converge to the state described in Condition **C1**.*

Under the control law (2.6), the vehicles are in constant motion with $e_t < 0$ (i.e., error is always decreasing) as long as the Condition **C1** is not satisfied. It utilizes the gradient of the error distribution inside $\mathscr{W}_i(t)$ to move in directions with maximum error. Hence it locally seeks to maximize coverage. However, using the control law (2.6) alone does not guarantee full coverage of at least C^* every where within $\mathscr{D}$. This is of no concern, since this lack of full effective coverage implies that $e \neq 0$, which will induce some vehicle to return and recover these partially covered regions. Hence, the following control strategy is proposed.

Control Strategy. Under the control law (2.6), all vehicles in the system are in continuous motion as long as the state described in Condition **C1** is avoided. Whenever the Condition **C1** holds with nonzero error $e(t) \neq 0$, the system has to be perturbed by switching to some other control law $\bar{\bar{\mathbf{u}}}_i$ that ensures violating Condition **C1**. Once away from Condition **C1**, the controller is switched back to the nominal control $\bar{\mathbf{u}}_i$ in Equation (2.6). Only when both Condition **C1** and $e(t) = 0$ are satisfied is when there is no need to switch to $\bar{\bar{\mathbf{u}}}_i$. Thus, the goal is to propose a simple linear feedback controller that guarantees driving the system away from Condition **C1**.

Now consider a simple perturbation control law that drives the system away from Condition **C1**. Define the time varying set:

$$\mathscr{D}_e(t) = \{\tilde{\mathbf{q}} \in \mathscr{D} : \mathscr{T}_{\mathscr{S}}(\tilde{\mathbf{q}},t) < C^*\}. \tag{2.7}$$

Let $\overline{\mathscr{D}}_e(t)$ be the closure of $\mathscr{D}_e(t)$. For each vehicle $\mathscr{V}_i$, let $\widetilde{\mathscr{D}}_e^i(t)$ denote the set of points in $\overline{\mathscr{D}}_e(t)$ that minimize the distance between $\mathbf{q}_i(t)$ and $\overline{\mathscr{D}}_e(t)$. That is,

$$\widetilde{\mathscr{D}}_e^i(t) = \left\{ \bar{\tilde{\mathbf{q}}} \in \overline{\mathscr{D}}_e(t) : \bar{\tilde{\mathbf{q}}} = \arg\min_{\tilde{\mathbf{q}} \in \overline{\mathscr{D}}_e(t)} \|\mathbf{q}_i(t) - \tilde{\mathbf{q}}\| \right\}.$$

This choice is efficient since the perturbation maneuver seeks the minimum-distance for redeployment.

Let t_s be the time at which Condition **C1** holds and $e(t_s) > 0$ while $e_t(t_s) = 0$. That is, t_s is the time of entry into the state described in Condition **C1** with nonzero error. At t_s, for each vehicle $\mathscr{V}_i$, consider a point $\tilde{\mathbf{q}}_i^*(t_s) \in \widetilde{\mathscr{D}}_e^i(t_s)$. Note that the set $\widetilde{\mathscr{D}}_e^i(t_s)$ may include more than a single point. Consider the control law

$$\bar{\bar{\mathbf{u}}}_i(t) = -\bar{\bar{k}}_i \left(\mathbf{q}_i(t) - \tilde{\mathbf{q}}_i^*(t_s) \right). \tag{2.8}$$

Under the regime when $e_t = 0$ and $e(t) > 0$, this control law is a simple linear feedback controller and will drive each vehicle in the fleet towards its associated $\tilde{\mathbf{q}}_i^*(t_s)$. Note that it is possible that $\tilde{\mathbf{q}}_i^*(t_s) = \tilde{\mathbf{q}}_j^*(t_s)$ for some pair $i \neq j \in \mathscr{S}$. By simple linear systems theory, the feedback control law (2.8) will result in having $\mathbf{q}_i(\hat{t}_s)$, for some $i \in \mathscr{S}$, be inside a ball of radius $\varepsilon < r_i$ at some time $\hat{t}_s > t_s$. Hence the point $\mathbf{q}_i^*$ is guaranteed to lie strictly inside the sensory range of vehicle $\mathscr{V}_i$.

The above discussion proves the following result.

Theorem 2.2.1. *Under sensor model properties **SM1-3** and **IC1**, the control law*

$$\mathbf{u}_i^*(t) = \begin{cases} \bar{\mathbf{u}}_i & \text{if Condition } \boldsymbol{C1} \text{ does not hold} \\ \bar{\bar{\mathbf{u}}}_i & \text{if Condition } \boldsymbol{C1} \text{ holds} \end{cases}, \tag{2.9}$$

drives the error $e(t) \to 0$ *as* $t \to \infty$.

2.2.3 Underwater Coverage Control with Vision-Based AUVs

In this section, the dynamic coverage control problem is utilized for underwater applications, such as sampling, surveillance, and search and rescue/retrieval, using a fleet of cooperative submarines. A sensor model based on a vision-based camera is presented.

Underwater exploration is important for many scientific and industrial applications. However, the literature on multi- and single-vehicle underwater application is relatively new due to recent advances in autonomous underwater vehicles (AUVs) and underwater positioning and communication technologies. Cooperative underwater MAVs have a wide range of applications that include sampling, oceanography, weather prediction [37, 75], studying aquatic life [56], mine countermeasure and hydrographic reconnaissance [129], search and rescue/retrieval [67], and archaeology [89]. Furthermore, due the the rapid attenuation of light and sound in sea water, advanced underwater survey technologies and AUV motion control strategies are of great interest.

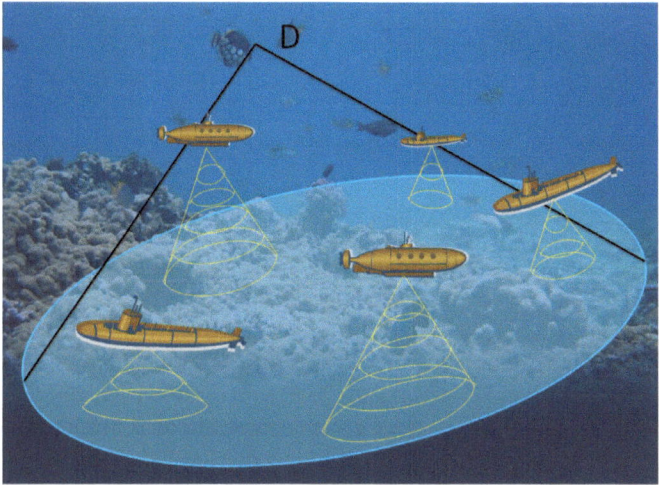

Fig. 2.2 Underwater sampling using a fleet of cooperative submarines.

2.2.3.1 Sensor Model

In underwater applications, domains of interest are generally three-dimensional (3D) volumes in the ocean with AUVs moving in all three directions. Figure 2.2 shows the scenario of underwater sampling using a fleet of cooperative submarines with cone-like vision-based cameras.

Unlike laser-based sensors and radars, vision-based camera sensors acquire data in a non-invasive way. They can be used for some specific applications for which visual information plays a basic role. This is especially suitable for underwater missions. Moreover, there is no interference among sensors of the same type, which could be critical for a large number of vehicles moving simultaneously in the same environment [11].

For the sake of simplicity, consider a simpler case where all the submarines move along a horizontal line (the configuration space Q) and thus the sampling domain $\mathscr{D}$ becomes an area (that is, a rectangle) below this line of motion (see Figure 2.3). In this scenario, each submarine looks in the vertically downward (or upward) direction. Therefore, domain $\mathscr{D}$ could also be above Q or both above and below Q, depending on which direction(s) the cameras are pointed. All vehicles are assumed linear kinematics given by Equation (2.1), where $\mathbf{u}_i = (u_{ix}, u_{iy}) \in \mathbb{R}^2$ is the control velocity of submarine $\mathscr{V}_i$. Note that there is no control in the vertical z direction other than control forces that maintain buoyancy of the submarine. The results obtained from this simple scenario can easily be generalized to two-dimensional horizontal motions. For the three-dimensional configuration space case, however, the gravitational and buoyancy forces in the vertical direction need to be included, which introduces nonlinearities in the equations of motion.

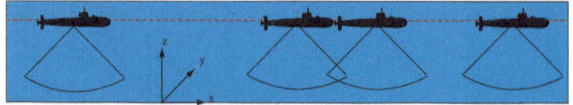

Fig. 2.3 One-dimensional configuration space scenario

In the general case where $\mathscr{D} \subset \mathbb{R}^3$, the sensor is a vision-based camera whose sensing domain is a three-dimensional cone. In the simpler case where $Q = \mathbb{R}^1$ and $\mathscr{D}$ is a compact subset of $\mathbb{R}^2$, the sensing domain becomes a sector, and one can use polar coordinates to define the instantaneous coverage function of the camera sensor model. For brevity and simplicity of exposure, the sensor model for the $\mathscr{D} \subset \mathbb{R}^2$ case is described in the following paragraphs. Extension to the three-dimensional cone model is easily performed by working with spherical coordinates (introducing an additional angular component) instead of polar coordinates.

In polar coordinates, let a point $\tilde{\mathbf{q}} \in \mathscr{D}$ be represented by (ρ_i, θ_i) with respect to submarine $\mathscr{V}_i$'s position $\mathbf{q}_i$. As shown in Figure 2.4, here the radial coordinate ρ_i represents the radial distance from the camera position $\mathbf{q}_i$ to $\tilde{\mathbf{q}}$, and θ_i is the counterclockwise angular coordinate angle from the vertical axis passing through the camera attached to submarine $\mathscr{V}_i$.

The sensor model **SM** in Section 2.2.1 is modified as follows: The sensing ability of each digital camera declines along the radial distance and the radial angle. That is, when the value of ρ_i and $|\theta_i|$ increases, the sensing ability of the camera decreases until it becomes zero at the maximum sensing range $\bar{\rho}$ and the maximum sensing direction Θ. Here the same maximum radial distance and maximum radial angle is assumed for each camera on all submarines as $\bar{\rho}$ and Θ. This is done without loss of generality as the ensuing results can be easily modified to reflect different maximum ranges and directions $\bar{\rho}_i$ and Θ_i, respectively. Hence, A_i is a function of both the radial distance ρ_i and radial angle angle θ_i. Mathematically, the sensory domain $\mathscr{W}_i$ of each submarine is given by

$$\mathscr{W}_i(t) = [\tilde{\mathbf{q}} \subset \mathscr{D} \mid \rho_i = \|\mathbf{q}_i(t) - \tilde{\mathbf{q}}\| \leq \bar{\rho} \text{ and}$$
$$|\theta_i| = \left| \arctan\left(\frac{q_{ix} - \tilde{q}_x}{q_{iz} - \tilde{q}_z} \right) \right| \leq \Theta \}. \tag{2.10}$$

The minimum sensing ability is given by

$$0 = A_i((q_{ix} + \rho_i \sin \Theta, q_{iz} + \rho_i \cos \Theta), (q_{ix}, q_{iz}))$$
$$= A_i((q_{ix} + \bar{\rho} \sin \theta_i, q_{iz} + \bar{\rho} \cos \theta_i), (q_{ix}, q_{iz})).$$

Note that the y and z components, q_{iy}, q_{iz}, of $\mathbf{q}_i$ are constant in the linear configuration space case.

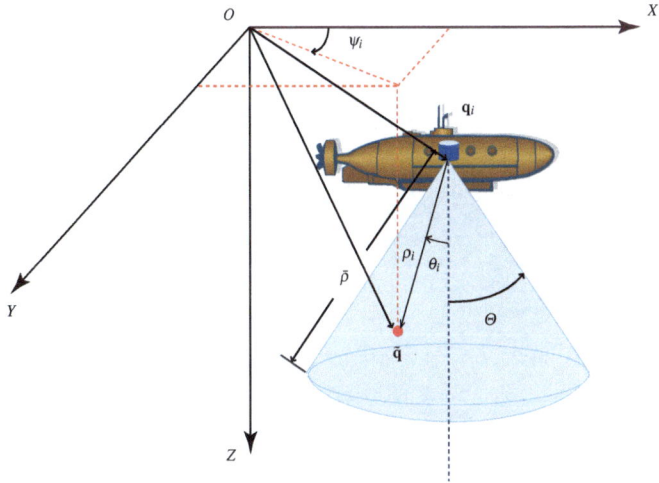

Fig. 2.4 Camera Sensor Model.

An example of a sector-like instantaneous coverage function A_i is a two-variable second order polynomial function of

$$d_i = \|\mathbf{q}_i - \tilde{\mathbf{q}}\|^2, \quad \alpha_i = \arctan^2\left(\frac{q_{ix} - \tilde{q}_x}{q_{iz} - \tilde{q}_z}\right).$$

within the sensor range and zero otherwise. In particular, consider the function

$$A_i(d_i, \alpha_i) = \begin{cases} \frac{M_i(d_i - R^2)^2(\alpha_i - \Theta^2)^2}{\bar{\rho}^4 \Theta^4} & \text{if } d_i \leq \bar{\rho}^2, \alpha_i \leq \Theta^2 \\ 0 & \text{otherwise} \end{cases} \tag{2.11}$$

An example for the instantaneous coverage function (2.11) is given by Figure 2.5 with $\mathbf{q}_i = 0, M_i = 1, \bar{\rho} = 12$, and $\Theta = \frac{5\pi}{18}$ in planar field.

A three-dimensional model can easily be obtained from this model by adding an additional angular variable ψ_i and restricting the angular extent of the model to some maximum value Ψ similar to the treatment of θ_i above.

Remark 2.2.1. *This sensor model is similar to the one which combines camera and ultrasonic sensor used in the YAMABICO robot [98]. For a vision-based sensor model applied to a 3D configuration space scenario with 9 viewpoints, see [18] for more details.* ●

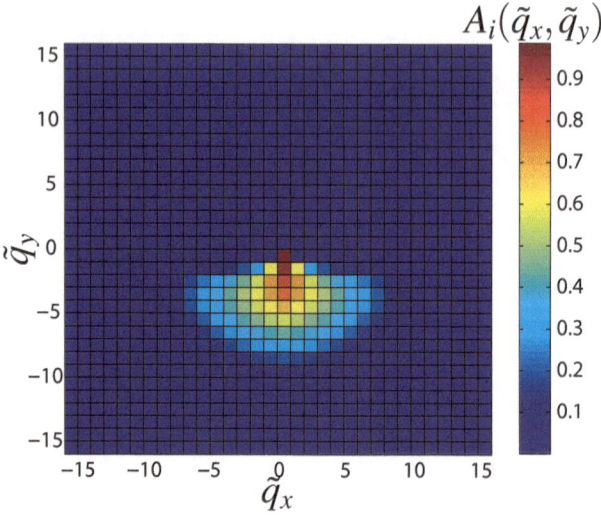

Fig. 2.5 Instantaneous coverage function for vision-based sensor model.

2.2.3.2   Control Law

In this section, the control law (2.6) in Section 2.2.2 is adopted and the domain search strategies are developed according to the modified vision-based sensor model presented in the above section.

Consider the following nominal control law[1]

$$\bar{\mathbf{u}}_i(t) = -\bar{k}_i \int_{\mathscr{D}} h'(C^* - \mathscr{T}_{\mathscr{S}}(\tilde{\mathbf{q}},t)) \left(\frac{\partial A_i}{\partial d_i} \frac{\partial d_i}{\partial \mathbf{q}_i} + \frac{\partial A_i}{\partial \alpha_i} \frac{\partial \alpha_i}{\partial \mathbf{q}_i} \right) \phi(\tilde{\mathbf{q}}) d\tilde{\mathbf{q}}, \qquad (2.12)$$

where $\bar{k}_i > 0$ are fixed feedback gains.

Using the same perturbation control law as Equation (2.8) and following similar derivation as in Section 2.2.2, a similar theorem as Theorem 2.2.1 can be derived. This guarantees that every point within the underwater search domain will be sampled by C^*. The performance of the proposed search strategy is demonstrated by the following simulation results.

---

[1] For brevity, the two-dimensional sensor model is assumed where A_i is a function of the two variables d_i and α_i. For the general three-dimensional conic sensor model case, the results can easily be extended by adding the additional term ψ_i in the model of A_i and adding one more term ($\frac{\partial A_i}{\partial \psi_i}$) when taking the derivatives of A_i.

2.2.3.3 Simulation

In this section a numerical simulation is provided to illustrate the performance of the coverage control strategy with the perturbation control law that ensures the global coverage.

As previously mentioned, the configuration space Q is a closed interval (all submarines move on a line). The domain $\mathscr{D}$ should be the area obtained by "extruding" the interval Q downwards to a depth of $\bar{\rho}$ (the maximum radial distance of the vision-based sensor). However, according to the sensor model, the sensing ability at the maximum radial distance is zero, which means that the domain to be covered has to be shallower than the distance $\bar{\rho}$ from where the submarines are located. Therefore, in the simulations shown here, the domain $\mathscr{D}$ is defined as a rectangle region whose size is $(\bar{\rho} - \bar{z}) \times l$ units length, where l is the length of the interval Q and $\bar{z} > 0$ is a fixed variable. The quantity $\bar{z}$ is chosen as 4 in the following simulations.

The maximum radius of the vision-based camera $\bar{\rho}$ is chosen as 12 and l is 40. There are 4 submarines ($N = 4$) with a randomly selected initial deployment as shown in Figure 2.6(a). Let the desired effective coverage C^* be 40. Here the control law in Equation (2.12) is used with control gains $\bar{k}_i = 1 \times 10^{-5}, i = 1, \ldots, 4$. Assume that there is no prior information as to the accuracy of the underwater sampling and, hence, $\phi(\breve{\mathbf{q}})$ is set as 1 for all $\breve{\mathbf{q}} \in \mathscr{D}$. For the sensor model, let $M_i = 1, \Theta = \frac{2\pi}{5}$ for all $i = 1, \ldots, 4$. A simple trapezoidal method is used to compute integration over $\mathscr{D}$ and a simple first order Euler scheme to integrate with respect to time.

The results are shown in Figures 2.6 and 2.7. Figure 2.6(a) shows the fleet motion along the line where each submarine is denoted by a different color. Figure 2.6(b) shows the control effort as a function of time. Figure 2.6(c) shows the global error $e(t)$ with switching control and can be seen to converge to zero. Figure 2.7 shows the effective coverage (dark blue for low and yellow for full coverage) and fleet configuration at $t = 0, 90, 180, 270, 360, 450$ with the perturbation control law.

2.2.4 Underwater Acoustic Imaging Using AUVs

This section studies the underwater acoustic imaging problem using AUVs. The integration of a guidance/control scheme and acoustic imaging process is discussed. A sensor model based on an acoustic sensor's beam pattern is presented. The goal is to obtain an accurate enough image of an underwater profile. Acoustic imaging is an active research field devoted to the study of techniques for the formation and processing of images generated from raw signals acquired by an acoustic system [93, 63].

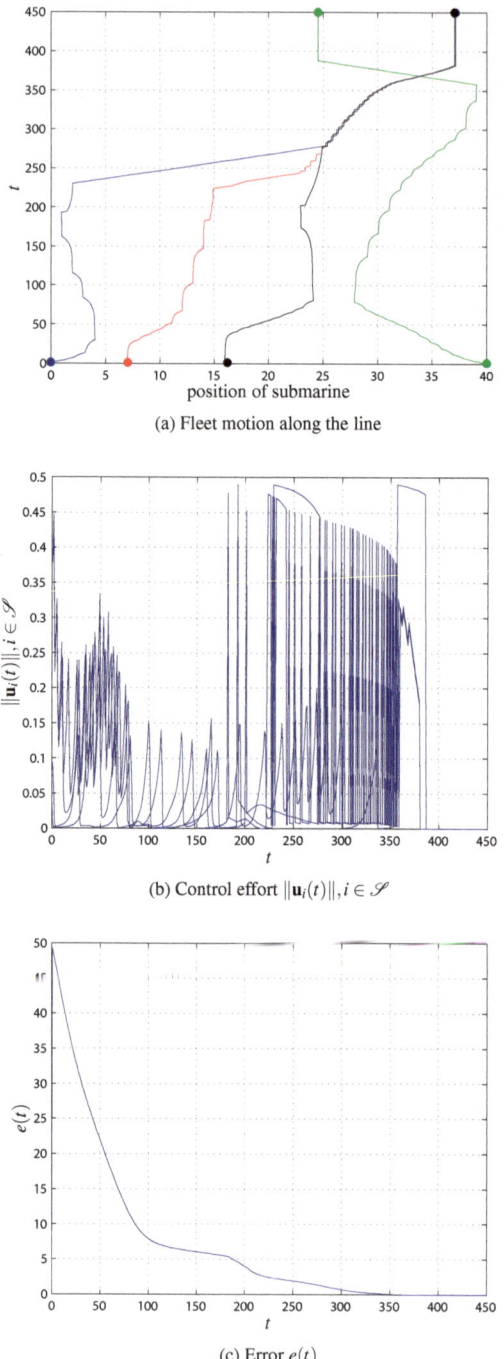

(a) Fleet motion along the line

(b) Control effort $\|\mathbf{u}_i(t)\|, i \in \mathscr{S}$

(c) Error $e(t)$

Fig. 2.6 Fleet motion, control velocity, and error for underwater applications.

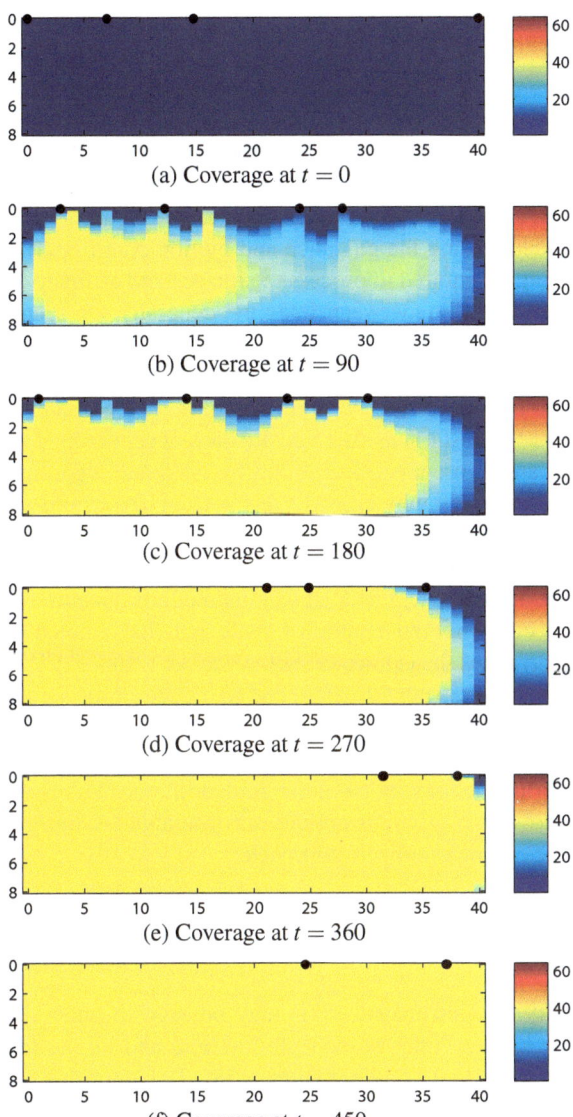

(a) Coverage at $t = 0$

(b) Coverage at $t = 90$

(c) Coverage at $t = 180$

(d) Coverage at $t = 270$

(e) Coverage at $t = 360$

(f) Coverage at $t = 450$

Fig. 2.7 Evolution of coverage with perturbation control.

2.2.4.1   Integration of Guidance, Control and Acoustic Imaging

The system is composed of two main tasks: vehicle motion guidance for coverage control and acoustic image processing for seafloor mapping.

The basic goal of the control part is to use a fleet of AUVs to collect enough imaging data at each location in an underwater domain. The assumptions of 1D configuration space and linear kinematics for the AUVs still hold here. While collecting imaging data during the guidance and motion control part, the technology of acoustic imaging is required to process the images and estimate the profile of the seabed simultaneously. In underwater imaging, generally, the scene under investigation, the seabed in our case, is first insonified by an acoustic signal $\mathscr{S}(t)$, then the backscattered echoes acquired by the system are processed to create the profile. This process can be performed by two different approaches: use of an acoustic lens followed by a retina of acoustic sensors, or acquisition of echoes by a two-dimensional array of sensors and subsequent processing by adequate algorithms, such as the beamforming or the holography class. In this section, the beamforming algorithm [63] is adopted to process the acoustic image. Each vehicle is mounted with a sensor array. It is assumed that an acoustic pulse $\mathscr{S}(t)$ is emitted and a spherical propagation occurs inside an isotropic, linear, absorbing medium. Beamforming is a spatial filter that linearly combines the temporal signals spatially sampled by the sensor array. The system arranges the echoes in such a way as to amplify the signal coming from a fixed direction (steering direction) and to reduce all the signals coming from the other directions. More details of the beamforming method will be presented in Section 2.2.4.2.

When considering the integration of the guidance/control scheme and the acoustic imaging process, two different options are available for the guidance system: either a stochastic or a deterministic approach.

Image Quality Feedback Based Error Guidance

The image quality (i.e., estimated error) may be used to guide the vehicles. For example, use the Kalman filter to estimate the field and on the filter's prediction step to solve for the vehicle's best next move [58]. The algorithm presented therein guarantees that the vehicles move to the direction that maximizes the quality of the estimated field.

Sensor Model Based Feedback Guidance

The sensor model (given by the beam pattern function, see next section) may also be used for vehicle guidance. In this section, this deterministic guidance approach will be adopted together with the beamforming algorithm.

2.2.4.2 Mathematical Summary of Acoustic Imaging

Beamforming Data Acquisition

Assume that the imaged scene is made up of m_s point scatterers, the i_{th} scatterer is placed at the position $\mathbf{r}_i = (x_i, z_i)$, as shown in Figure 2.8. Define the plane $z = 0$ as the plane that receives the backscattered field. The acoustic signal $\mathscr{S}(t)$ is emitted by an ideal point source placed in the coordinate origin (i.e., at vehicle location). Consider N_s point like sensors that constitute a receiving 2-D array, numbered by index l, from 0 to $N_s - 1$. The steering direction of a beam signal is then indicated by the angle ϑ measured with respect to the z axis. By applying the Fourier/Fresnel approximation, one can obtain the following expression for the beam signal:

$$b(t, \vartheta) = \sum_{i=1}^{m_s} \mathscr{S}(t - \frac{2\rho_i}{c}) C_i \mathrm{BP}_{\mathrm{BMF}}(\omega, \theta_i, \vartheta), \tag{2.13}$$

$$\mathrm{BP}_{\mathrm{BMF}}(\omega, \theta, \vartheta) = \frac{sin[\omega N_s d(sin\theta - sin\vartheta)/2c]}{sin[\omega d(sin\theta - sin\vartheta)/2c]}, \tag{2.14}$$

where C_i is some constant related to the i_{th} scatterer, c is the speed of sound, $\mathrm{BP}_{\mathrm{BMF}}(\omega, \theta, \vartheta)$ is called beam pattern, which depends on the arrival angle θ, the steering angle ϑ, and the angular frequency ω. It is also assumed that the array is equispaced and centered in the coordinate origin, and d is the inter-element spacing. Figures 2.9(a) and 2.9(b) show the beam pattern for a $N_s = 40$ element array with $d = 1.5$mm spacing as a function of the arrival angle θ (visualized on a logarithmic scale normalized to 0 dB) for fixed frequency $f = 500$KHz and steering angle $\vartheta = 0°$C and $\vartheta = 30°$C, respectively[63].

Imaging Processing

The analysis of beam signals allows one to estimate the range to a scene. A common method to detect the distance of the scattering object is to look for the maximum peak of the beam signal envelope. Denoting by t^* the time instant at which the maximum peak (whose magnitude is denoted by s^*) occurs, the related distance, R^*, is easily derivable from it (i.e., $R^* = c \cdot t^*/2$, if the pulse source is placed in the coordinate origin). Therefore, for each steering direction ϑ, a triplet (ϑ, R^*, s^*) can be extracted. The set of triplets can be projected to get a range image in which the point defined in polar coordinates by ϑ and R^* is converted into a Cartesian point (x^*, z^*) [94].

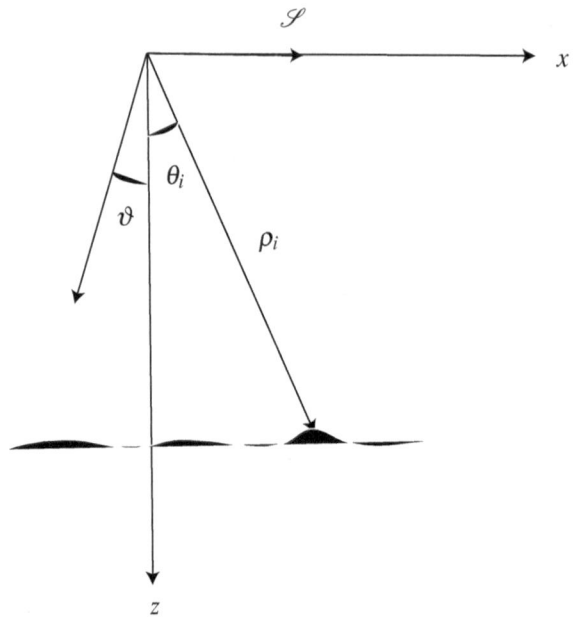

Fig. 2.8 Geometry of the data model.

2.2.4.3 Control Law

The beam pattern BP given by Equation (2.14) is used as a sensor model to describes how effective the vehicle surveys a point $\tilde{\mathbf{q}} \in \mathscr{D}$. The effective coverage of the group indexed by $\mathscr{K}$ at time t at the point $\tilde{\mathbf{q}}$ becomes:

$$\mathscr{T}_{\mathscr{K}}(\tilde{\mathbf{q}},t) = \int_0^t \sum_{i \in \mathscr{K}} \mathrm{BP}_i^2(\tau) d\tau$$

Assume BP_i is a function of θ_i here only, that is, the steering direction ϑ and angular frequency ω are fixed. Since BP_i is a function of θ_i which varies with time because of the change of vehicle position, BP_i is implicitly a function of time.

Consider the following nominal control law

$$\bar{\mathbf{u}}_i(t) = \bar{k}_i \int_{\mathscr{D}} h'(C^* - \mathscr{T}_{\mathscr{S}}(\tilde{\mathbf{q}},t)) \left(\sum_{i \in \mathscr{S}} \mathrm{BP}_i \right) \frac{\partial \mathrm{BP}_i}{\partial \theta_i} \frac{\partial \theta_i}{\partial \mathbf{q}_i} \phi(\tilde{\mathbf{q}}) d\tilde{\mathbf{q}}, \qquad (2.15)$$

where $\bar{k}_i > 0$ are fixed feedback gains.

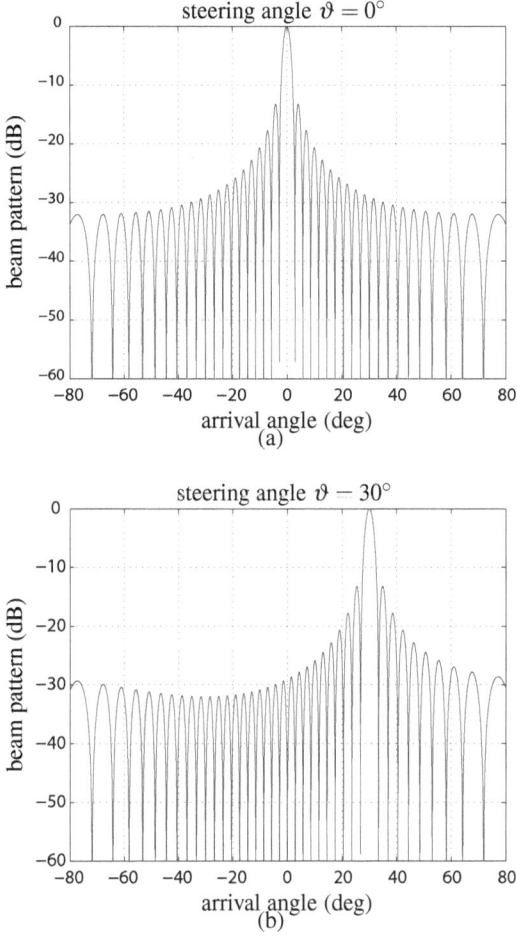

Fig. 2.9 Beam pattern.

Together with the perturbation control law $\bar{\bar{\mathbf{u}}}_i(t)$ given by Equation (2.8), the overall control strategy guarantees full coverage of every point within the domain. This can be proved following a same fashion as Theorem 2.2.1.

2.2.4.4 Simulation

This section provides a set of numerical simulations. Define the length of $\mathscr{D}$ as $l = 20$ meters in the following simulation. The seabed profile is given by a simple piecewise linear function

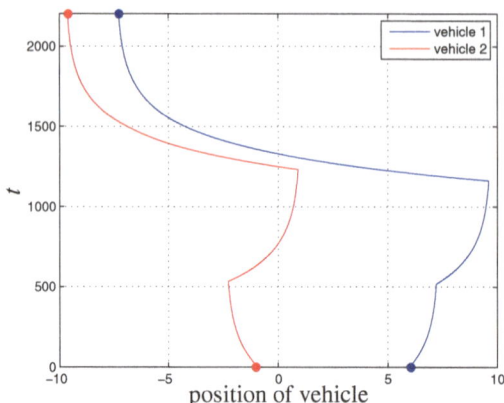

Fig. 2.10 Fleet motion along the line in acoustic imaging.

$$y = \begin{cases} -gx \text{ if } x \leq 0 \\ gx \quad \text{if } x > 0 \end{cases}$$

where x is the discretization along the seabed length and $g = 2.5$ is the slope of the linear function. Assume there are 2 submarines ($N = 2$) with a randomly selected initial deployment as shown in Figure 2.10. Let the desired effective coverage C^* be 6000. Here the control law in Equation (2.15) is used with control gains $\bar{k}_i = 0.05, i = 1,2$. For the beam pattern sensor model, set $f = 500$kHz, $\vartheta = 0$, $d = 1.5$mm, $N_s = 40$, $c = 1500$m/s for all $i = 1,2$. The sensor has a Gaussian random noise with zero mean and a standard deviation of 0.5.

The control effort $\|\mathbf{u}_i\|$, $i \in \mathscr{S}$ is shown in Figure 2.11. The global error $e(t)$ is shown in Figure 2.12. It can be seen to converge to zero. Note that the error is normalized by dividing $(C^*)^2 \times l$ so that the initial error is 1. Figure 2.13 shows the effective coverage at $t = 367,734,2152$ with perturbation control laws.

The acoustic image measured by the vehicles using the algorithm discussed in Section 2.2.4.2 is shown in Figure 2.14. It compares the actual seabed profile with the simulated curve. The result shows that even with sensor noise, the proposed algorithm efficiently estimates the actual profile.

2.3 Deterministic Awareness-Based Approach

In the previous section, a Lyapunov-based coverage control strategy is proposed to guarantee the completion of a domain search mission under a deterministic framework. Remaining in the deterministic framework, in this section, an awareness-based dynamic model is developed, which describes how "aware" a system of networked, limited-range MAVs is of events occurring at every point over a given domain. The approach aims at modeling the dynamic information loss over time within

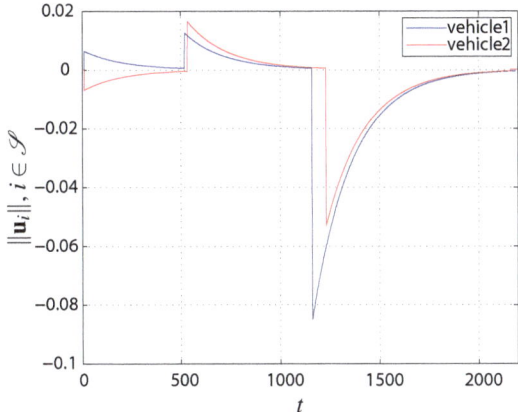

Fig. 2.11 Vehicle control effort in acoustic imaging.

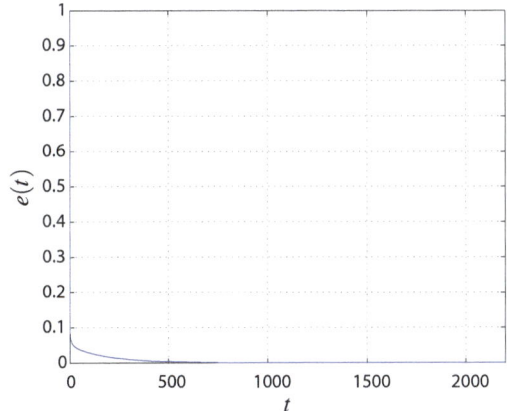

Fig. 2.12 Global error $e(t)$ in acoustic imaging.

the search domain. This formulation can be applied to a wide variety of problems, including large-scale and complex domains, that may be disconnected (surveillance over adversarial pockets in a region), or hybrid discrete and continuous (surveillance over urban environments and inside buildings, where roads and hallways are the continuous part of the domain, and buildings and rooms are discrete nodes).

The proposed awareness model will be first applied to the coverage control over large-scale task domains using decentralized MAVs with intermittent communications and/or faulty sensors. For each vehicle, an individual state of awareness is defined. The individual vehicle's state of awareness continuously evolves based on the vehicle's motion and is updated at discrete instants whenever the vehicle establishes a communication link with other vehicles. This information sharing update step aids in reducing the amount of redundant coverage. The hybrid nature of the

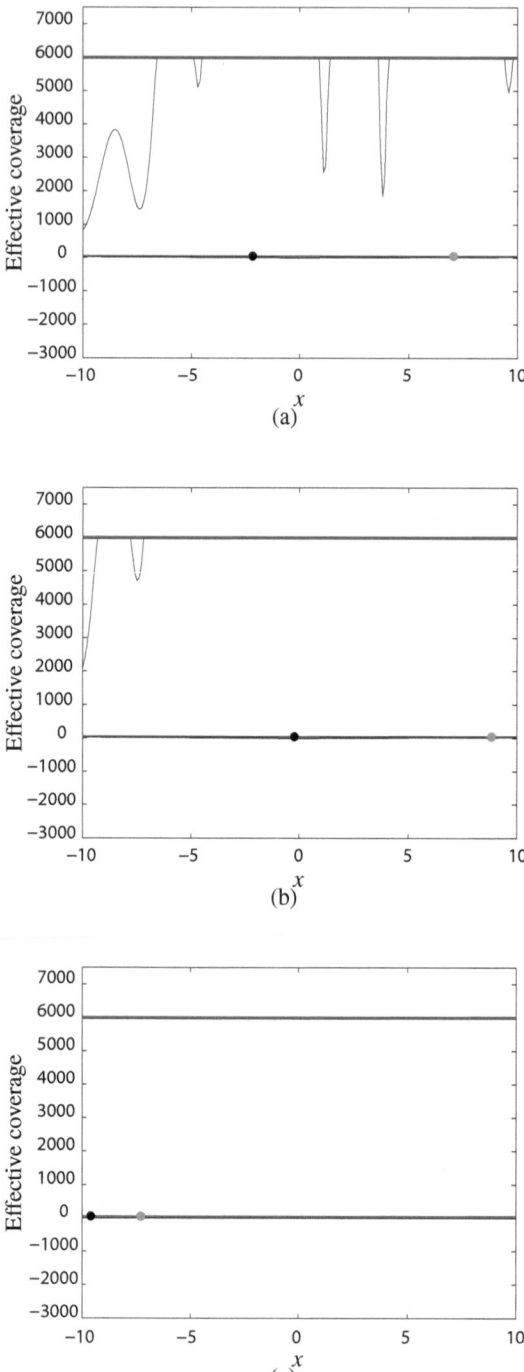

Fig. 2.13 Effective coverage in acoustic imaging.

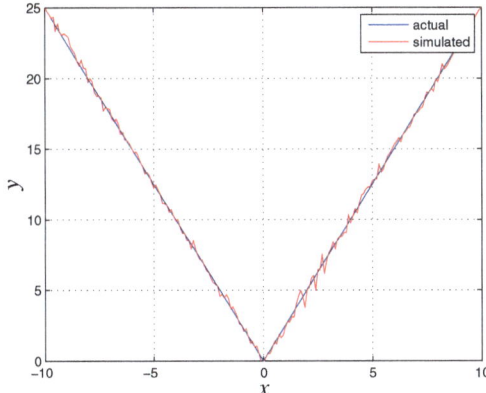

Fig. 2.14 Actual versus simulated profile

"awareness" dynamic model and the intermittent communications between the vehicles result in a switching closed-loop control law. Based on this awareness model, a decentralized control strategy is proposed that guarantees that every point within the task domain will be covered with a satisfactory state of awareness under intermittent communications and/or faulty sensors.

The intermittent communication structure is desirable because in most cases it is not energy efficient or even possible for the vehicle fleet to maintain open communication channels during the entire mission. This is especially true for large-scale task domains, where vehicles may need to disperse (and, hence, lose connectivity with other vehicles) in order to cover the domain. In [121], smooth control laws using potential functions are developed for stable flocking motion of mobile agents. A similar flocking problem is studied in [122] and [123] under a connected (but with arbitrary dynamic switching) decentralized networks. Both discrete-time and continuous-time consensus update schemes are proposed in [107] for distributed multi-agent systems in the presence of switching interaction topologies. In [5], a distributed Kalman consensus algorithm is proven to converge to an unbiased estimate for both static and dynamic communication networks. In [143], the authors investigate distributed mobile robots in a wireless network under nearest neighbor communications. In [83], local undirected communication is used in fully distributed multi-agent systems. Both [143] and [83] demonstrate improvements in global behavior made by exchanging local sensing information.

For the sake of completeness, the dynamics of an individual state of awareness is generalized to the total awareness achieved by a fleet of MAVs. The corresponding centralized search strategies are proposed where all the vehicles share awareness information.

2.3.1   Problem Formulation

A description of large-scale domains has already been given in Chapter 1. Here, a rigorous mathematical definition is given as follows: A large-scale domain is one where, under the best case scenario when all the sensory domains $\mathscr{W}_i$ are disjoint, there exists a set $\varXi \subset \mathscr{D}$ of non-zero measure such that for every $\tilde{\mathbf{q}} \in \varXi$, $\tilde{\mathbf{q}} \notin \mathscr{W}_i$ for all $i \in \mathscr{S}$. Note that the results derived in the following sections also apply to non-large-scale domains. The first-order kinematic equation of motion (2.1) and the sensor model **SM** (2.3) are assumed for each vehicle. The limited-range sensor models the practical difficulty in real implementation, especially for missions over large-scale domains.

State of Awareness

An individual vehicle's state of awareness is a distribution $\tilde{\mathbf{x}}_i(\tilde{\mathbf{q}},t) : \mathbb{R}^2 \times \mathbb{R} \to \mathbb{R}$ that is a measure of how "aware" the vehicle $\mathscr{V}_i$ is of events occurring at a specific location $\tilde{\mathbf{q}}$ at time t. Here, without loss of generality, assume that $\tilde{\mathbf{x}}_i(\tilde{\mathbf{q}},t) \in [0,1]$, that the initial state of awareness is zero (i.e., no awareness), and that the desired state of awareness is given by 1 (full awareness), while $\tilde{\mathbf{x}}_i(\tilde{\mathbf{q}},t) < 1$ corresponds to insufficient awareness. Fixing a point $\tilde{\mathbf{q}} \in \mathscr{D}$, the state of awareness of a particular vehicle $\mathscr{V}_i$ at time t is assumed to satisfy the following differential equation

$$\dot{\tilde{\mathbf{x}}}_i(\tilde{\mathbf{q}},t) = -\left(A_i(\|\mathbf{q}_i - \tilde{\mathbf{q}}\|) - \zeta\right)\left(\tilde{\mathbf{x}}_i(\tilde{\mathbf{q}},t) - 1\right), \ \tilde{\mathbf{x}}_i(\tilde{\mathbf{q}},0) = \tilde{\mathbf{x}}_{i0} = 0, \ i \in \mathscr{S}, \quad (2.16)$$

where $\zeta \geq 0$ is a constant parameter which models the loss of awareness over a time period during which no vehicles cover a point of interest in $\mathscr{D}$. Having $\zeta > 0$, sets a periodical re-visit requirement to maintain desired awareness levels.

Let $\mathbf{x}_i(\tilde{\mathbf{q}},t) = \tilde{\mathbf{x}}_i(\tilde{\mathbf{q}},t) - 1$ be the transformed state of awareness. The dynamics of the transformed state of awareness is then given by

$$\dot{\mathbf{x}}_i(\tilde{\mathbf{q}},t) = -\left(A_i(\|\mathbf{q}_i - \tilde{\mathbf{q}}\|) - \zeta\right)\mathbf{x}_i(\tilde{\mathbf{q}},t), \ \mathbf{x}_i(\tilde{\mathbf{q}},0) = \mathbf{x}_{i0} = -1, \ i \in \mathscr{S}. \quad (2.17)$$

Therefore, the transformed state of awareness $\mathbf{x}_i(\tilde{\mathbf{q}},t) \in [-1,0]$. The initial transformed state of awareness is -1, which reflects the fact that at the outset of the surveillance mission the fleet has poor awareness levels. One may set a nonuniform initial distribution for $\mathbf{x}_i(\tilde{\mathbf{q}},t)$ to reflect any prior awareness knowledge. The nonuniform distribution $\mathbf{x}_i(\tilde{\mathbf{q}},0)$ may reflect regions where objects may be able to camouflage themselves better than in other regions of $\mathscr{D}$ (e.g., dense forests versus open fields). A more negative value of $\mathbf{x}_i(\tilde{\mathbf{q}},0)$ reflects areas with less awareness levels, and vice versa. However, the initial value is always restricted to be greater than -1, with -1 representing the worst case scenario (which is the assumption made here).

For the transformed state of awareness, the desired equilibrium awareness level is zero, that is, $\mathbf{x}_i(\tilde{\mathbf{q}},t) = 0$, $t > 0$, $i \in \mathscr{S}$, $\forall \tilde{\mathbf{q}} \in \mathscr{D}$.

A control law will be developed to guarantee the convergence of $\mathbf{x}_i(\tilde{\mathbf{q}},t)$ to a neighborhood of 0: $\|\mathbf{x}_i(\tilde{\mathbf{q}},t)\| < \xi$ for some $\xi > 0$, which corresponds to $\tilde{\mathbf{x}}_i(\tilde{\mathbf{q}},t)$ approaching unity and a state of full domain awareness. Note that under the dynamics (2.17), the maximum value attainable by $\mathbf{x}_i(\tilde{\mathbf{q}},t)$ is zero if the initial awareness level is negative. Inspecting Equation (2.17), the system state of awareness is degrading except over regions where $A_i - \zeta$ has a positive value (i.e., $0 \le \zeta \le A_i$).

One can also define the overall transformed awareness dynamics:

$$\dot{\mathbf{x}}(\tilde{\mathbf{q}},t) = -\sum_{i=1}^{N} \left(A_i(\|\mathbf{q}_i - \tilde{\mathbf{q}}\|) - \zeta\right)\mathbf{x}(\tilde{\mathbf{q}},t) \tag{2.18}$$

with negative initial conditions as discussed above for $\mathbf{x}_i(\tilde{\mathbf{q}},t)$. Here $\sum_{i=1}^{N} A_i(\|\mathbf{q}_i - \tilde{\mathbf{q}}\|)$ is the total instantaneous coverage achieved by all the vehicles at time t. The overall awareness dynamics will be utilized to develop the centralized search control laws. If one wishes to consider the state of awareness achieved by a set $\mathcal{K} \subset \mathcal{S}$, then one can use Equation (2.18) but summing only over elements in $\mathcal{K}$. Note that $\mathbf{x}_i \le \mathbf{x}$. That is, the overall awareness of the sensors in a centralized system is better than that of the individual sensors in a decentralized system. Note that for the case where all the vehicles are set to be fixed, if enough resources are available (i.e., enough vehicles and/or large enough sensor ranges) the entire domain can be covered with $\sum_{i=1}^{N}(A_i - \zeta) > 0$ and the awareness level is everywhere increasing and converging to the desired value: $\mathbf{x}(\tilde{\mathbf{q}},t) \to 0$ for all $\tilde{\mathbf{q}} \in \mathcal{D}$. This is guaranteed to occur using a static sensor network and a sufficiently small domain $\mathcal{D}$ (the small-scale domain case). This is true because for each point $\tilde{\mathbf{q}}$, the term $\sum_{i=1}^{N}(A_i - \zeta)$ in Equation (2.18) is a positive constant since each vehicle is assumed to be fixed. This means that, for each $\tilde{\mathbf{q}}$, the dynamics (2.18) is a linear differential equation in $\mathbf{x}(\tilde{\mathbf{q}},t)$, which leads to asymptotic convergence of $\mathbf{x}(\tilde{\mathbf{q}},t)$ to zero. For large-scale domains, a static sensor is guaranteed not to meet the desired zero transformed state of awareness because, by definition, there exists a set of non-zero measure $\Xi \subset \mathcal{D}$ which is not covered by some sensor. It is aimed to develop a decentralized control strategy that stabilizes the state of awareness under intermittent communications and/or faulty sensors over a large-scale domain.

Remark 2.3.1. *Let $\mathbf{x}_i(\tilde{\mathbf{q}},t) = 0$ for all $\tilde{\mathbf{q}} \notin \mathcal{D}$ and all $t \ge 0$. This remark will be useful for the validation of a lemma developed later.* ●

2.3.2 State of Awareness Dynamic Model

State of Awareness Updates

Consider the case where the vehicles communicate only when they are within a range $\lambda > 0$ of each other. If a communication channel is established, vehicles exchange their awareness information. Let $\mathcal{G}_i(t) = \{j \in \mathcal{S} : \|\mathbf{q}_j - \mathbf{q}_i\| < \lambda\}, i \in \mathcal{S}$, be the set of vehicles that neighbor vehicle $\mathcal{V}_i$ (including vehicle $\mathcal{V}_i$ itself) at time t.

Whenever new vehicles $\mathscr{V}_j$ are added to the set $\mathscr{G}_i$, vehicle $\mathscr{V}_i$ will instantaneously exchange all the available awareness information with new neighbors in a discrete awareness update step. If no or more than one vehicle drop from $\mathscr{G}_i(t)$ (possibly faulty sensors), the individual state of awareness of vehicle $\mathscr{V}_i$ does not change. Let t_c be the time instant at which vehicles $\mathscr{V}_j, \mathscr{V}_k, \dots$ become members of $\mathscr{G}_i$. That is $\mathscr{V}_j, \mathscr{V}_k \dots \notin \mathscr{G}_i(t_c^-)$ but $\mathscr{V}_j, \mathscr{V}_k \dots \in \mathscr{G}_i(t_c^+)$. Hence, the following update equation takes place whenever a set of vehicles $\bar{\mathscr{G}}_i(t) \subset \mathscr{S} \setminus \mathscr{G}_i(t)$ gets added to $\mathscr{G}_i(t)$ at time t:

$$\mathbf{x}_i(\tilde{\mathbf{q}}, t^+) = (-1)^{\bar{n}_i(t)} \mathbf{x}_i(\tilde{\mathbf{q}}, t) \cdot \prod_{j \in \bar{\mathscr{G}}_i(t)} \mathbf{x}_j(\tilde{\mathbf{q}}, t), \tag{2.19}$$

where $\bar{n}_i(t)$ is the number of vehicles in $\bar{\mathscr{G}}_i(t)$. Hence, the transformed state of awareness evolves according to the continuous dynamics given by Equation (2.17) and undergoes a discrete update step given by Equation (2.19) whenever new vehicles become $\mathscr{V}_i$'s neighbors. Figure 2.15 illustrates the awareness model for the continuous dynamics (2.17) and the discrete awareness state update (2.19). Note that there is one continuous mode (2.17) and one switching condition $\bar{\mathscr{G}}_i(t) \neq \varnothing$. When the switching condition is satisfied, the initial condition of the system is reset according to the reset map (2.19). If $\bar{\mathscr{G}}_i(t) = \varnothing$ (i.e., no new vehicles become neighbors of $\mathscr{V}_i$), then the awareness state of vehicle $\mathscr{V}_i$ obeys the continuous differential equation (2.17). This includes the case when vehicles drop from $\mathscr{G}_i(t)$ (e.g., faulty sensors) or when existing neighbors retain their $\mathscr{V}_i$ neighborhood status. If the number of new vehicles $\bar{n}_i(t)$ in $\bar{\mathscr{G}}_i(t)$ is nonzero at time t, then the value of the transformed state of awareness of vehicle $\mathscr{V}_i$ will be discretely substituted with the product of the awareness states of all the vehicles in $\bar{\mathscr{G}}_i(t)$ and itself. According to Equation (2.19), if the number of newly added vehicles is even, then the multiplication of their states of awareness will be a non-negative number because the term $\mathbf{x}_i$ is always less than or equal to zero. In this case, the newly updated state of awareness will stay negative after multiplying the state of awareness of vehicle $\mathscr{V}_i$ itself. However, when the number of newly added vehicles is odd, the multiplication of all these states of awareness together with the state of awareness of vehicle $\mathscr{V}_i$ will be a positive number. Hence, the introduction of $(-1)^{\bar{n}_i(t)}$ makes sure that the updated state of awareness is always negative. Moreover, this product reflects the improvement in the state of awareness of vehicle $\mathscr{V}_i$. For example, assume that all the vehicles in the mission fleet have an initial transformed state of awareness of -1 and their coverage goal is to achieve a transformed awareness value close to zero everywhere within the domain. If $\mathscr{V}_i$ has a transformed awareness of -0.5 at some $\tilde{\mathbf{q}}$ at time t, and it updates its transformed state of awareness based on the transformed state of awareness of another neighbor vehicle of -0.5, then the new awareness at $\tilde{\mathbf{q}}$ is now -0.25 according to the update Equation (2.19). The two extremes are:

1. if the second vehicle has no awareness at $\tilde{\mathbf{q}}$ (i.e., a value of -1), then the new awareness is still -0.5 since the second vehicle did not "add any awareness" at that point.

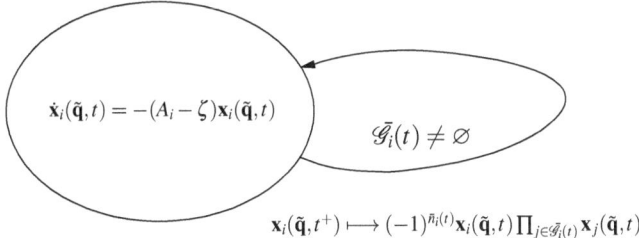

$$x_i(\tilde{\mathbf{q}}, t^+) \longmapsto (-1)^{\bar{n}_i(t)} x_i(\tilde{\mathbf{q}}, t) \prod_{j \in \bar{\mathscr{G}}_i(t)} x_j(\tilde{\mathbf{q}}, t)$$

Fig. 2.15 Continuous and discrete awareness state update model.

2. if the second vehicle has perfect awareness at $\tilde{\mathbf{q}}$ (i.e., a value of 0), then the new awareness is now 0 since the second vehicle had perfect awareness level there.

Awareness Metric

Let the awareness metric be given by

$$e_{gi}(t) = \int_{\mathscr{D}} \frac{1}{2} x_i^2(\tilde{\mathbf{q}}, t) d\tilde{\mathbf{q}}, \ i \in \mathscr{S}, \tag{2.20}$$

which is the global error over the entire mission domain achieved by vehicle $\mathscr{V}_i$. It is said to be global since the integration is performed over the entire domain $\mathscr{D}$. The coverage goal of each vehicle is to guarantee that the above metric (2.20) decreases with time and ultimately converges to a small neighborhood of zero.

Further, define

$$e_g(t) = \int_{\mathscr{D}} \frac{1}{2} \mathbf{x}^2(\tilde{\mathbf{q}}, t) d\tilde{\mathbf{q}} \tag{2.21}$$

as the global error over the entire mission domain achieved by all the MAVs.

Let the local awareness error function associated with vehicle $\mathscr{V}_i$ be

$$e^i_{\mathscr{W}_i}(t) = \int_{\mathscr{W}_i(t)} \frac{1}{2} x_i^2(\tilde{\mathbf{q}}, t) d\tilde{\mathbf{q}} \geq 0, \ i \in \mathscr{S}, \tag{2.22}$$

with $e^i_{\mathscr{W}_i}(t) = 0$ if and only if $x_i(\tilde{\mathbf{q}}, t) = 0$ for every point $\tilde{\mathbf{q}}$ inside the sensory domain $\mathscr{W}_i(t)$. This is a decentralized awareness metric associated with vehicle $\mathscr{V}_i$ that reflects the quality of the state of awareness within $\mathscr{W}_i(t)$ achieved by vehicle $\mathscr{V}_i$ alone. This metric will be used for the development of the decentralized control law. Note that the metric is a function of the position of vehicle $\mathscr{V}_i$ because of the integral domain $\mathscr{W}_i(t)$.

Moreover, define the centralized awareness metric associated with the entire search fleet $\mathscr{S}(t)$ by

$$e_{\mathscr{W}_i}(t) = \int_{\mathscr{W}_i(t)} \frac{1}{2} \mathbf{x}^2(\tilde{\mathbf{q}}, t) \mathrm{d}\tilde{\mathbf{q}}. \tag{2.23}$$

This is a centralized awareness metric associated with vehicle $\mathscr{V}_i$ that reflects the quality of the state of awareness within $\mathscr{W}_i(t)$ achieved by all vehicles in $\mathscr{S}$. This metric will be used to develop the control law for the centralized search problem.

2.3.3   Awareness Coverage with Intermittent Communications

Overall Description of Control Strategy

In this section, a decentralized control law $\mathbf{u}_i^*$ is developed based on the awareness metric (2.20) and the local awareness error function (2.22) over a large-scale domain using MAVs with intermittent communications and/or faulty sensors. The control law $\mathbf{u}_i^*$ is inspired by Equation (2.9) in Section 2.2.2 for deterministic Lyapunov-based coverage control. It is composed of a nominal control law $\bar{\mathbf{u}}_i$ and a perturbation control law $\bar{\bar{\mathbf{u}}}_i$. Initially, a vehicle $\mathscr{V}_i$ is deployed and is governed by a nominal control law $\bar{\mathbf{u}}_i$ developed based on the error function (2.22), which drives it in the direction that maximizes its local state of awareness (since the error function (2.22) is defined within the sensory domain $\mathscr{W}_i(t)$) by moving in the direction of low awareness levels. The nominal control law $\bar{\mathbf{u}}_i$ will eventually drive $e_{\mathscr{W}_i}^i(t)$ to a neighborhood of zero. Whenever the transformed state of awareness is such that $\|\mathbf{x}_i(\tilde{\mathbf{q}}, t)\| \leq \xi$, where ξ is some threshold to be defined later, for all $\tilde{\mathbf{q}} \in \mathscr{W}_i(t)$ (i.e., $e_{\mathscr{W}_i}^i(t) \to 0$), the vehicle is said to have converged to a local minimum, and the control law is switched to a perturbation control law $\bar{\bar{\mathbf{u}}}_i$ that drives the vehicle out of this local minimum to the nearest point with less than full awareness, which guarantees that every point within the domain $\mathscr{D}$ with insufficient awareness will be covered. Once away from the local minimum, $e_{\mathscr{W}_i}^i(t)$ is no longer in a small neighborhood of zero since not every point within the sensory domain $\mathscr{W}_i(t)$ has $\|\mathbf{x}_i(\tilde{\mathbf{q}}, t)\| \leq \xi$, and the controller is switched back to the nominal controller. The switching between the nominal control law $\bar{\mathbf{u}}_i$ and the perturbation control law $\bar{\bar{\mathbf{u}}}_i$ is repeated until the entire domain $\mathscr{D}$ has a full state of awareness. That is, the global error $e_{gi}(t)$ given by Equation (2.20) converges to a neighborhood of zero. Figure 2.16 illustrates the overall control strategy applied on a single vehicle for the awareness coverage control over a square domain. Green represents low awareness, yellow for higher awareness and red for full awareness. The black dot represents the position of the vehicle, while the circle indicates the limited range of the sensor. Figure 2.16(a) shows an initial deployment of the vehicle under the nominal control law $\bar{\mathbf{u}}_i$ at the outset. The control law $\bar{\mathbf{u}}_i$ moves the vehicle towards the direction of lower awareness levels. Figure 2.16(b) demonstrates an instance when the vehicle is trapped in a local minimum with full awareness and the perturbation control law $\bar{\bar{\mathbf{u}}}_i$ is applied. Figure 2.16(c) corresponds to full awareness, i.e., the mission is completed when $e_{gi} \longrightarrow 0$.

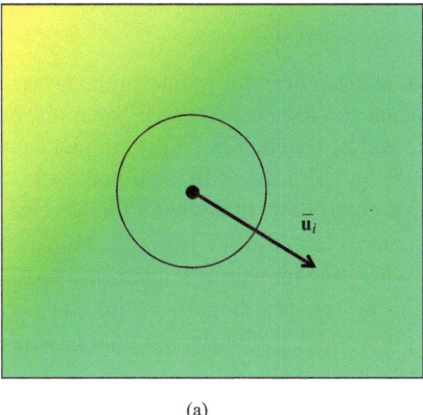

(a)

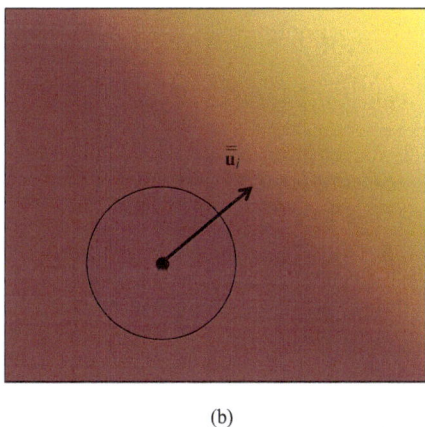

(b)

(c)

Fig. 2.16 Illustration of the overall control strategy.

Nominal Control Law

Between discrete jumps in awareness due to intermittent sharing of awareness information with other vehicles, the vehicle kinematic equation (2.1) and state of awareness equation (2.17) constitute two first order differential equations. In this section, these two equations together with the individual vehicle error function (2.22) are used to derive a nominal control law that seeks to reduce the value of $e^i_{\mathscr{W}_i}$ for each vehicle. The nominal control law itself does not guarantee convergence of $\mathbf{x}_i(\tilde{\mathbf{q}}, t)$ to a neighborhood of zero over the entire domain $\mathscr{D}$. Instead, it only guarantees that $\mathbf{x}_i(\tilde{\mathbf{q}}, t) \to 0$ within the sensory domain $\mathscr{W}_i$ for each vehicle. A perturbation control law will be deployed along with the nominal control law to guarantee that $\|\mathbf{x}_i(\tilde{\mathbf{q}}, t)\| < \xi$ over the entire domain $\mathscr{D}$.

Without any loss of generality, the following assumption for the initial state of awareness will be made.

IC2 The initial state of awareness is given by:

$$\mathbf{x}_i(\tilde{\mathbf{q}}, 0) = \mathbf{x}_{i0} = -1, i \in \mathscr{S}.$$

Assumption 2.3.1. *Assume that* $\zeta = 0$.

With Assumption 2.3.1, the ensuing results are applicable to problems in search and rescue/retrieval problems (especially with static victims or objects of interests), domain monitoring, and "low level" surveillance.

Consider the following nominal control law:

$$\bar{\mathbf{u}}_i(t) = \bar{k}_i \int_{\mathscr{W}_i(t)} \mathbf{x}_i^2(\tilde{\mathbf{q}}, t) \underbrace{\left(\int_0^t \frac{\partial (A_i(\tilde{\mathbf{q}}, \mathbf{q}_i(\sigma)))}{\partial \tilde{\mathbf{q}}} d\sigma \right)}_{\text{memory term}} d\tilde{\mathbf{q}}, \qquad (2.24)$$

where $\bar{k}_i > 0$ is a feedback gain. It will be proved that control law (2.24) guarantees the convergence of $\mathbf{x}_i(\tilde{\mathbf{q}}, t)$ to zero at every point $\tilde{\mathbf{q}}$ in the sensory domain $\mathscr{W}_i(t)$.

Remark 2.3.2. *In the expression for* $\bar{\mathbf{u}}_i(t)$*, the time integral "memory" term under the spatial integration is an integration of historical data that translates into the reliance on past search history for vehicle motion planning. Note that the memory term is multiplied by* $\mathbf{x}_i^2(\tilde{\mathbf{q}}, t)$ *before being integrated over the sensory domain at the current time t. This indicates that historical data as well as up-to-date awareness levels within the vehicle's sensor domain are compounded to decide on the motion.* •

First consider the following lemma, which will be used shortly.

Lemma 2.3.1. *For any function* $F : \mathbb{R}^2 \times \mathbb{R} \to \mathbb{R}$*, the following equation holds,*

$$\frac{d}{dt} \int_{\mathscr{W}_i(t)} F(\tilde{\mathbf{q}}, t) d\tilde{\mathbf{q}} = \int_{\mathscr{W}_i(t)} \left[(grad_{\tilde{q}} F(\tilde{\mathbf{q}}, t)) \cdot \mathbf{u}_i + \frac{\partial F(\tilde{\mathbf{q}}, t)}{\partial t} \right] d\tilde{\mathbf{q}},$$

where $\mathbf{u}_i$ *is the velocity of vehicle* $\mathscr{V}_i$ *and* $grad_{\tilde{\mathbf{q}}}$ *is the gradient operator with respect to* $\tilde{\mathbf{q}}$.

Proof. This is a direct consequence of Equation (3.3) in [38], where note that $\mathbf{u}_i$ is the velocity of any point within the (rigid) domain $\mathscr{W}_i$ (including the boundary). ☐

Next, consider the following condition, whose utility will also become obvious shortly.

Condition C2. $\mathbf{x}_i(\tilde{\mathbf{q}},t) = 0, \forall \tilde{\mathbf{q}} \in \mathscr{W}_i(t)$.

This condition corresponds to the case where the set of points within $\mathscr{W}_i$ have perfect coverage and the local error $e^i_{\mathscr{W}_i}$ is zero.

Lemma 2.3.2. *For any* $t \geq 0$, *if Condition **C2** holds for vehicle* $\mathscr{V}_i$, *then* $e^i_{\mathscr{W}_i}(t) = 0, i \in \mathscr{S}$. *Conversely, if* $e^i_{\mathscr{W}_i}(t) = 0$ *for some time* $t \geq 0$, *then Condition **C2** holds for vehicle* $\mathscr{V}_i$.

Proof. The proof of this lemma is similar to that of Lemma 2.2.1 in Section 2.2.2. ☐

Theorem 2.3.1. *Under Assumption 2.3.1, the control law* $\bar{\mathbf{u}}_i(t)$ *given by Equation (2.24) drives* $e^i_{\mathscr{W}_i}(t) \longrightarrow 0$ *asymptotically between awareness state switches.*

Proof. Consider the function $\bar{V}_i = e^i_{\mathscr{W}_i}(t) \geq 0$. From Lemma 2.3.2, $\bar{V}_i = 0$ if and only if Condition **C2** holds for vehicle $\mathscr{V}_i$. According to Lemma 2.3.1,

$$\dot{\bar{V}}_i = \dot{e}^i_{\mathscr{W}_i}(t) = \frac{d}{dt} \int_{\mathscr{W}_i(t)} \frac{1}{2} \mathbf{x}_i^2(\tilde{\mathbf{q}},t) d\tilde{\mathbf{q}}$$

$$= \int_{\mathscr{W}_i(t)} grad(\frac{1}{2} \mathbf{x}_i^2(\tilde{\mathbf{q}},t)) \cdot \bar{\mathbf{u}}_i d\tilde{\mathbf{q}} + \int_{\mathscr{W}_i(t)} \frac{\partial(\frac{1}{2} \mathbf{x}_i^2(\tilde{\mathbf{q}},t))}{\partial t} d\tilde{\mathbf{q}}. \quad (2.25)$$

Note that according to Remark 2.3.1, the integration region $\mathscr{W}_i(t)$ always holds even when $\mathscr{W}_i(t)$ lies outside of $\mathscr{D}$. First consider the spatial gradient term in Equation (2.25):

$$\int_{\mathscr{W}_i(t)} grad(\frac{1}{2} \mathbf{x}_i^2(\tilde{\mathbf{q}},t)) \cdot \bar{\mathbf{u}}_i d\tilde{\mathbf{q}} = \int_{\mathscr{W}_i(t)} \frac{\partial(\frac{1}{2} \mathbf{x}_i^2(\tilde{\mathbf{q}},t))}{\partial \tilde{\mathbf{q}}} \cdot \bar{\mathbf{u}}_i d\tilde{\mathbf{q}}$$

$$= \int_{\mathscr{W}_i(t)} \mathbf{x}_i(\tilde{\mathbf{q}},t) \frac{\partial(\mathbf{x}_i(\tilde{\mathbf{q}},t))}{\partial \tilde{\mathbf{q}}} \cdot \bar{\mathbf{u}}_i d\tilde{\mathbf{q}}.$$

Next, an expression for $\frac{\partial(\mathbf{x}_i(\tilde{\mathbf{q}},t))}{\partial \tilde{\mathbf{q}}}$ needs to be derived. From Equation (2.17) and assuming $\zeta = 0$, it follows that

$$\mathbf{x}_i(\tilde{\mathbf{q}},t) = e^{-\int_0^t A_i(\tilde{\mathbf{q}},\mathbf{q}_i(\sigma))d\sigma} \mathbf{x}_{i0}.$$

Hence,

$$\frac{\partial \mathbf{x}_i(\tilde{\mathbf{q}},t)}{\partial \tilde{\mathbf{q}}} = -e^{-\int_0^t A_i(\tilde{\mathbf{q}},\mathbf{q}_i(\sigma))d\sigma} \mathbf{x}_{i0} \int_0^t \frac{\partial (A_i(\tilde{\mathbf{q}},\mathbf{q}_i(\sigma)))}{\partial \tilde{\mathbf{q}}} d\sigma$$

$$= -\mathbf{x}_i(\tilde{\mathbf{q}},t) \int_0^t \frac{\partial (A_i(\tilde{\mathbf{q}},\mathbf{q}_i(\sigma)))}{\partial \tilde{\mathbf{q}}} d\sigma.$$

Therefore,

$$\int_{\mathscr{W}_i(t)} \text{grad}(\frac{1}{2}\mathbf{x}_i^2(\tilde{\mathbf{q}},t)) \cdot \bar{\mathbf{u}}_i d\tilde{\mathbf{q}} = -\int_{\mathscr{W}_i(t)} \mathbf{x}_i^2(\tilde{\mathbf{q}},t) \left(\int_0^t \frac{\partial (A_i(\tilde{\mathbf{q}},\mathbf{q}_i(\sigma)))}{\partial \tilde{\mathbf{q}}} d\sigma \right) \cdot \bar{\mathbf{u}}_i d\tilde{\mathbf{q}}.$$

Note that $\bar{\mathbf{u}}_i(t)$ is a function of time but not $\tilde{\mathbf{q}}$, so it can be pulled outside of the integration. Substitute $\bar{\mathbf{u}}_i(t)$ in Equation (2.24) into the above equation, it follows that

$$\int_{\mathscr{W}_i(t)} \text{grad}(\frac{1}{2}\mathbf{x}_i^2(\tilde{\mathbf{q}},t)) \cdot \bar{\mathbf{u}}_i d\tilde{\mathbf{q}} = -\bar{k}_i \left[\int_{\mathscr{W}_i(t)} \mathbf{x}_i^2(\tilde{\mathbf{q}},t) \left(\int_0^t \frac{\partial (A_i(\tilde{\mathbf{q}},\mathbf{q}_i(\sigma)))}{\partial \tilde{\mathbf{q}}} d\sigma \right) d\tilde{\mathbf{q}} \right]^2 \leq 0.$$

Next, consider the integral of the time derivation term in Equation (2.25). According to Equation (2.17) and assuming no information loss, that is, $\zeta = 0$,

$$\int_{\mathscr{W}_i(t)} \frac{\partial (\frac{1}{2}\mathbf{x}_i^2(\tilde{\mathbf{q}},t))}{\partial t} d\tilde{\mathbf{q}} = -\int_{\mathscr{W}_i(t)} \mathbf{x}_i^2(\tilde{\mathbf{q}},t) A_i(\|\tilde{\mathbf{q}} - \mathbf{q}_i\|) d\tilde{\mathbf{q}} \leq 0.$$

Therefore,

$$\dot{V}_i = \underbrace{-\bar{k}_i \left[\int_{\mathscr{W}_i(t)} \mathbf{x}_i^2(\tilde{\mathbf{q}},t) \left(\int_0^t \frac{\partial (A_i(\tilde{\mathbf{q}},\mathbf{q}_i(\sigma)))}{\partial \tilde{\mathbf{q}}} d\sigma \right) d\tilde{\mathbf{q}} \right]^2}_{\text{first term}}$$

$$\underbrace{-\int_{\mathscr{W}_i(t)} \mathbf{x}_i^2(\tilde{\mathbf{q}},t) A_i(\|\tilde{\mathbf{q}} - \mathbf{q}_i\|) d\tilde{\mathbf{q}}}_{\text{second term}} \leq 0.$$

Note that equality holds if and only if Condition **C2** holds. This can be seen as follows. First, note that if Condition **C2** holds, $\dot{V}_i$ is clearly equal to zero because $\mathbf{x}_i(\tilde{\mathbf{q}},t) = 0$ for $\forall \tilde{\mathbf{q}} \in \mathscr{W}_i(t)$. Secondly, if $\dot{V}_i$ is zero, but $\mathbf{x}_i(\tilde{\mathbf{q}},t) \neq 0$ within $\mathscr{W}_i(t)$, the second term will always be non-zero because $A_i(\|\tilde{\mathbf{q}} - \mathbf{q}_i\|) > 0$ within the sensory domain $\mathscr{W}_i(t)$. Hence, for $\dot{V}_i = 0$, the only possibility is that $\mathbf{x}_i(\tilde{\mathbf{q}},t) = 0$, which also makes the first term zero. Then $\dot{V}_i = 0$ only if Condition **C2** holds. This and Lemma 2.3.2 complete the proof. $\square$

Perturbation Control Law

Before introducing the perturbation control law, consider the following condition.

Condition C3. $\|\mathbf{x}_i(\tilde{\mathbf{q}},t)\| \leq \xi, \forall \tilde{\mathbf{q}} \in \mathscr{W}_i(t)$, where $\xi > 0$ is the awareness tolerance.

This condition corresponds to the case where the local error (i.e., over $\mathscr{W}_i$) is in a neighborhood of zero, that is, the situation when the vehicle is making very little progress (almost "stuck").

Using the nominal control law in Equation (2.24), each vehicle will be guaranteed to have a state of awareness $\|\mathbf{x}_i(\tilde{\mathbf{q}},t)\| \leq \xi$ at each point $\tilde{\mathbf{q}} \in \mathscr{W}_i(t)$ for a given $\xi > 0$, i.e., Condition **C3**. However, this does not necessarily mean that the error $e_{gi}(t)$ of each vehicle over the entire domain given by Equation (2.20) will converge to a neighborhood of zero. If Condition **C3** holds but with $e_{gi}(t) > \bar{\xi}$ (to be precisely defined), the perturbation control law given by Equation (2.8) in Section 2.2.2 is used to perturb the system away from the Condition **C3**, however, here $\tilde{\mathbf{q}}_i^* \in \mathscr{D}$ is chosen such that $\|\mathbf{x}_i(\tilde{\mathbf{q}}_i^*,t_s)\| > \xi$. Define the following sets in a same fashion as in Section 2.2.2, i.e.,

$$\mathscr{D}_e^i(t) := \{\tilde{\mathbf{q}} \in \mathscr{D} : \|\mathbf{x}_i(\tilde{\mathbf{q}},t)\| > \xi\},$$

let $\overline{\mathscr{D}}_e^i(t)$ be the closure of $\mathscr{D}_e^i(t)$ and we have,

$$\widetilde{\mathscr{D}}_e^i(t) = \left\{\bar{\tilde{\mathbf{q}}} \in \overline{\mathscr{D}}_e^i(t) : \bar{\tilde{\mathbf{q}}} = \mathrm{argmin}_{\tilde{\mathbf{q}} \in \overline{\mathscr{D}}_e^i(t)} \|\mathbf{q}_i(t) - \tilde{\mathbf{q}}\|\right\}.$$

Here the superscripts i is used to indicate that the sets are associated with vehicle $\mathscr{V}_i$. Note that $\tilde{\mathbf{q}}_i^*$ is chosen based on coverage information available to vehicle $\mathscr{V}_i$ only, which is appropriate in the setting here since the control law is decentralized.

Theorem 2.3.2. *If the system is at the state described by the Condition* **C3** *and the set* $\widetilde{\mathscr{D}}_e^i(t)$ *at time t is nonempty, then the control law* $\bar{\mathbf{u}}_i(t)$ *given by Equation (2.8) drives the system away from Condition* **C3**.

Proof. If Condition **C3** holds and the set $\widetilde{\mathscr{D}}_e^i(t)$ at time t is nonempty, it follows from the linearity of the closed-loop system: $\dot{\mathbf{q}}_i(t) = -\bar{k}_i(\mathbf{q}_i(t) - \tilde{\mathbf{q}}_i^*(t_s))$ that the vehicle $\mathscr{V}_i$ will converge asymptotically to a neighborhood of $\tilde{\mathbf{q}}_i^*(t_s)$. Hence, there will exist a time such that $\|\mathbf{q}_i - \tilde{\mathbf{q}}_i^*\| < r_i$, at which time Condition **C3** no longer holds. $\square$

Overall Control Strategy

Theorems 2.3.1 and 2.3.2 give us the following result.

Theorem 2.3.3. *Under limited sensory range model* **SM** *and initial condition* **IC2**, *the control law*

$$\mathbf{u}_i^*(t) = \begin{cases} \bar{\mathbf{u}}_i & \text{if Condition } \textbf{C3} \text{ does not hold} \\ \bar{\bar{\mathbf{u}}}_i & \text{if Condition } \textbf{C3} \text{ holds} \end{cases}, \tag{2.26}$$

drives the error $e_{gi}(t), i \in \mathscr{S}$, *to a neighborhood of zero value.*

Proof. Under the control law (2.24), each vehicle moves in the direction that improves its own local (since integration is performed over the sensor domain $\mathscr{W}_i(t)$)

awareness level and is in continuous motion as long as the state described in Condition **C3** is avoided. Whenever the Condition **C3** holds with global error $e_{gi}(t) > \bar{\xi}, i \in \mathscr{S}$, the system is perturbed away from the Condition **C3** by switching to the perturbation control law (2.8). Once away from the Condition **C3**, the controller is switched back to the nominal controller. This procedure is repeated until the point in time when there does not exist $\tilde{\mathbf{q}}_i^*$ whenever Condition **C3** holds. The non-existence of such a $\tilde{\mathbf{q}}_i^*$ guarantees that $e_{gi}(t)$ is sufficiently close to zero (since $\|\mathbf{x}_i(\tilde{\mathbf{q}},t)\|$ is not larger than ξ everywhere). That is to say, only when both Condition **C3** holds and $\|\mathbf{x}_i(\tilde{\mathbf{q}},t)\| \leq \xi$, $(\|\tilde{\mathbf{x}}_i(\tilde{\mathbf{q}},t)\| \rightarrow 1)$, for all $\tilde{\mathbf{q}} \in \mathscr{D}$, the mission is said to be accomplished and no further switching is performed.

To complete the proof, one has to show that infinite switching between (1) the continuous awareness evolution (2.17) and discrete awareness update (2.19), and (2) the nominal control law $\bar{\mathbf{u}}_i$ (2.24) and the perturbation control law $\bar{\bar{\mathbf{u}}}_i$ (2.8) can never happen. For the former, note the fact that when $\mathbf{x}_i(\tilde{\mathbf{q}},t)$ undergoes a discrete update step, no instabilities are introduced. This is true since the update equation results in a discrete change from a continuous distribution $\mathbf{x}_i(\tilde{\mathbf{q}},t)$ over $\mathscr{D}$ to another continuous distribution $\mathbf{x}_i(\tilde{\mathbf{q}},t^+)$. Moreover, $\|\mathbf{x}_i(\tilde{\mathbf{q}},t^+)\| \leq \|\mathbf{x}_i(\tilde{\mathbf{q}},t)\|$ for each $\tilde{\mathbf{q}}$ at each switching instant. Hence, the resetting of $\mathbf{x}_i(\tilde{\mathbf{q}},t)$ can not introduce unbounded divergence by design and can only result in the decrease in the norm of $\mathbf{x}_i(\tilde{\mathbf{q}},t)$.

Secondly, infinite switching between $\bar{\mathbf{u}}_i$ and $\bar{\bar{\mathbf{u}}}_i$ is impossible because (a) during the application of $\bar{\mathbf{u}}_i$ the value of e_{gi} decreases by an amount of non-zero measure, and (b) if Condition **C3** occurs and the control law $\bar{\bar{\mathbf{u}}}_i$ is applied, once the vehicle is within a range less than r_i from $\tilde{\mathbf{q}}_i^*$, e_{gi} decreases by an amount of non-zero measure. These two facts guarantee that a finite number of switches will be performed to reach $e_{gi} \leq \bar{\xi}$, where $\bar{\xi}$ is an upper bound given by

$$e_{gi}(t) = \int_{\mathscr{D}} \frac{1}{2}\mathbf{x}_i^2(\tilde{\mathbf{q}},t)\mathrm{d}t = \left\|\int_{\mathscr{D}} \frac{1}{2}\mathbf{x}_i^2(\tilde{\mathbf{q}},t)\mathrm{d}t\right\| = \int_{\mathscr{D}} \frac{1}{2}\left\|\mathbf{x}_i^2(\tilde{\mathbf{q}},t)\right\|\mathrm{d}t \leq \frac{\xi^2 A_{\mathscr{D}}}{2} = \bar{\xi},$$

where $A_{\mathscr{D}}$ is the area of $\mathscr{D}$.

Finally, it also needs to show that the control velocity $\mathbf{u}_i^*(t)$ can never be infinite. When $\mathbf{x}_i(\tilde{\mathbf{q}},t)$ undergoes resetting, the control law $\bar{\mathbf{u}}_i$ undergoes a finite drop in magnitude (since $\mathbf{x}_i^2(\tilde{\mathbf{q}},t)$ itself experiences a finite drop in magnitude, see Equation (2.24), and since the memory term indicated in Equation (2.24) does not change across switches) and, hence, no infinite control inputs are encountered across awareness state switches. In between switches, the control $\bar{\mathbf{u}}_i$ is also finite (but, in this case, continuous in time) because the memory term is finite (since the coverage function A_i is at least $\mathscr{C}^1$) for any finite time and since $\|\mathbf{x}_i(\tilde{\mathbf{q}},t)\|$ converges to a neighborhood of zero. The perturbation control law $\bar{\bar{\mathbf{u}}}_i(t)$ is clearly bounded in magnitude since the feedback gain is finite and the vector $\mathbf{q}_i(t) - \mathbf{q}_i^*(t_s)$ has a finite magnitude (due to boundedness of $\mathscr{D}$). ☐

Remark 2.3.3. *The search approach proposed herein requires computations at the order of $\mathscr{O}(\bar{n}^2+2)$ at each time step, where $\bar{n}$ is the number of cells in the discretized sensory domain $\mathscr{W}_i$. While alternative approaches, such as Voronoi-partitioning and* *stochastic-based SLAM methods, are computationally more burdensome (Refer to [60] for more details).*   ●

Remark 2.3.4. *As a matter of implementation, if the condition for the reset map and the Condition **C3** occur at the same instant, checking of the Condition **C3** is performed after the reset map is performed.*   ●

Remark 2.3.5. *Note that $\bar{\mathbf{u}}_i$ relies on the properties of the sensor coverage function A_i. Hence, the coverage control law relies on the given sensor model to guide the vehicle during the coverage mission.*   ●

Remark 2.3.6. *Redundant coverage (overlapping paths) would be expected among the vehicles. The main reasons for the overlapping of paths are:*

- *Decentralization and the fact that communications are established only intermittently, meaning that a vehicle may not have the actual overall history of coverage information. A main difference between the approach introduced here and [59] or [24] is that every sensor only considers a local error function $e^i_{\mathcal{W}_i}(t)$ that is independent of what other sensors may do. In other words, the sensor metrics presented here are independent of each other and do not capture the property of "cooperation in sensing." That is, cooperation is established in terms of interchange of information through communication only.*
- *Sometimes a vehicle has to traverse an already covered region in order to get to an uncovered region.*   ●

2.3.4 Generalization to Centralized Coverage Control

In this section, the above decentralized coverage control laws for MAVs are generalized to centralized coverage control laws, where the awareness information is shared over all vehicles in $\mathscr{S}$. Consider the following conditions.

IC3 The initial state of awareness is given by: $\mathbf{x}(\tilde{\mathbf{q}},0) = \mathbf{x}_0 = -1$.

Condition C4. $\mathbf{x}(\tilde{\mathbf{q}},t) = 0, \forall \tilde{\mathbf{q}} \in \mathscr{W}_i(t)$.

Here, the dynamics of $\mathbf{x}(\tilde{\mathbf{q}},t)$ follows Equation (2.18). Following similar procedures as above, the global awareness metric (2.21) and the local awareness error function (2.23) based on all the MAVs are utilized to develop a centralized control law $\mathbf{u}_i^*$.

$$\mathbf{u}_i^*(t) = \begin{cases} \bar{\mathbf{u}}_i(t) \text{ if Condition } \mathbf{C4} \text{ doesn't hold for } \mathscr{V}_i \in \mathscr{S} \\ \bar{\bar{\mathbf{u}}}_i(t) \text{ if Condition } \mathbf{C4} \text{ holds for } \mathscr{V}_i \in \mathscr{S} \end{cases} \qquad (2.27)$$

where

$$\bar{\mathbf{u}}_i(t) = \bar{k}_i \int_{\mathscr{W}_i(t)} \mathbf{x}^2(\tilde{\mathbf{q}},t) \underbrace{\left(\int_0^t \frac{\partial A_i(\tilde{\mathbf{q}}, \mathbf{q}_i(\sigma))}{\partial \tilde{\mathbf{q}}} d\sigma \right)}_{\text{memory term}} d\tilde{\mathbf{q}} \qquad (2.28)$$

is the centralized nominal control law, and the choice of the point $\tilde{\mathbf{q}}_i^*$ in the centralized perturbation control law is based on centralized awareness information such that $\|\mathbf{x}(\tilde{\mathbf{q}}, t)\| > \xi$.

Theorem 2.3.4. *Under limited sensory range model SM and initial condition IC3, the control law $\mathbf{u}_i^*(t)$ given by Equation (2.27) drives the error $e_g(t)$ to a neighborhood of zero value.*

This theorem can be proved following similar derivations as Theorem 2.3.3 without difficulty.

2.3.5 Simulation

In this section a numerical simulation is provided to illustrate the performance of the control strategy (2.26). Define the domain $\mathscr{D}$ as a square region whose size is 64×64 units length and thus naturally discretize it into 64×64 cells, where $\tilde{\mathbf{q}}$ represents the centroid of each cell. The domain has no information loss, that is, $\zeta = 0$. Assume there are 4 vehicles ($N = 4$) with a randomly selected initial deployment as shown by the green dots in Figure 2.17(a). Figure 2.17(a) shows the fleet motion in the plane (start at green dot and end at red dot). Let the initial state $\mathbf{x}_{i0}, i = 1, 2, 3, 4$, be -1 and the desired state for $\mathbf{x}_i(\tilde{\mathbf{q}}, t)$ be 0, which correspond to $\tilde{\mathbf{x}}_{i0} = 0$ and the desired actual state of awareness $\tilde{\mathbf{x}}_i(\tilde{\mathbf{q}}, t) = 1$. Here the nominal control law in Equation (2.24) is used with control gain $\bar{k}_i = 8$ and the perturbation control law in Equation (2.8) is used with control gain $\bar{\bar{k}}_i = 1, i = 1, \ldots, 4$. A vehicle is set to switch to the linear feedback control law whenever Condition **C3** applies to it with $\xi = 1e^{-3}$. For the sensor model, set $M_i = 1, r_i = 12$ for all $i = 1, \ldots, 4$. For the intermittent communication range, it is set as the same as the sensory range $\lambda = r_i = 12$. The control velocities for all vehicles are shown in Figure 2.17(b). The global error $e_g(t)$ plotted in Figure 2.17(c) is the actual total performance achieved by the entire vehicle fleet and can be seen to converge to zero.

Figure 2.18 shows the variation of the transformed state of awareness $\mathbf{x}(\tilde{\mathbf{q}}, t)$ during the coverage mission, which is the equivalent awareness level as achieved by all the MAVs. Note that the minimal transformed state of awareness is about -5.2×10^{-3} over the entire domain at $t = 325$ and that the global error metric converges to a neighborhood of zero as predicted by Theorem 2.3.3.

2.4 Probabilistic Bayesian-Based Approach

In this section, the coverage control strategies are developed under a probabilistic framework, which guarantee full certainty over the mission domain based on Bayes analysis and information theory. In practice, no matter how high the quality of a vehicle sensor is, its sensing capability is limited and erroneous observations

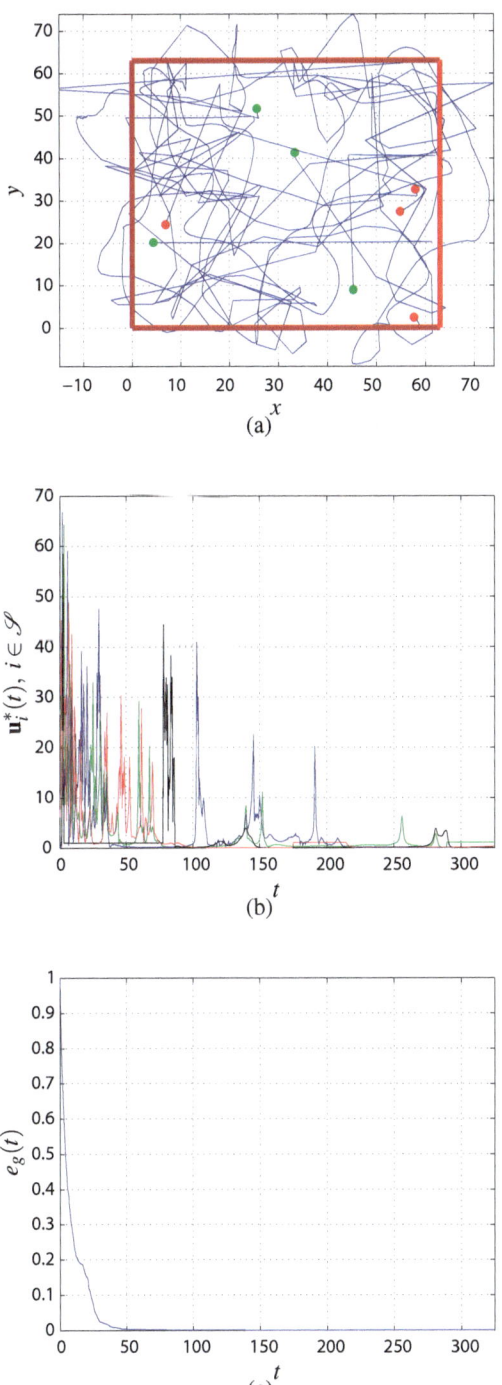

Fig. 2.17 Fleet motion, control effort, and error for awareness coverage control.

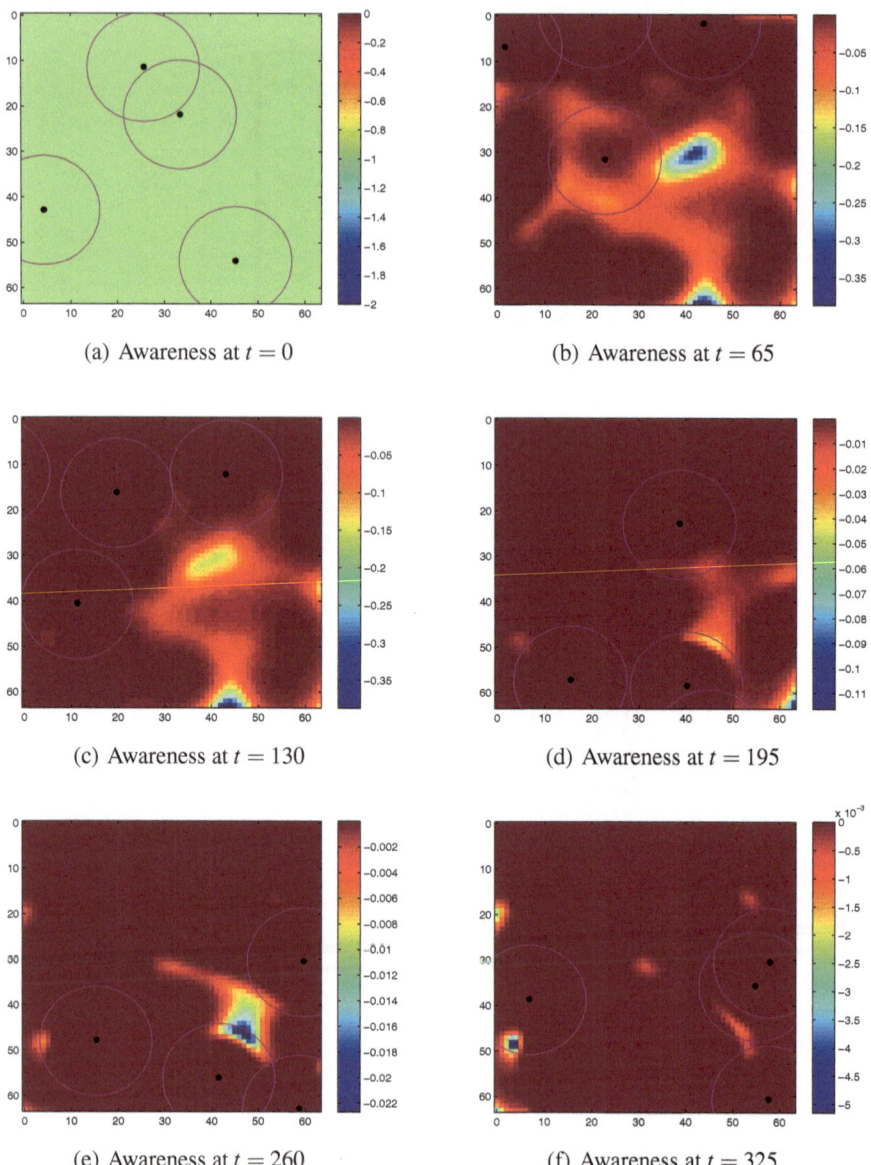

(a) Awareness at $t = 0$

(b) Awareness at $t = 65$

(c) Awareness at $t = 130$

(d) Awareness at $t = 195$

(e) Awareness at $t = 260$

(f) Awareness at $t = 325$

Fig. 2.18 Transformed state of awareness $\mathbf{x}(\tilde{\mathbf{q}}, t)$.

are bound to occur due to noise and sensor failure [124]. Hence, false or missed detections of object existence are inevitable and the system performance is indeterministic. Therefore, a probabilistic framework is desirable as it takes into account sensor errors, as well as allows for future incorporation of other tasks such as object tracking, data association, data/decision fusion, sensor registration, and clutter resolution.

In the stochastic setting, Bayes filters are used extensively for dynamic surveillance of a search domain. In [58], the author uses the Kalman filter for estimating a spatially-decoupled (i.e., it does not satisfy a partial differential equation, or a PDE) field and using the prediction step of the filter for guiding the vehicles to move in directions that improve the field estimate. The control algorithm is modified to guarantee satisfactory global coverage of the domain. Other stochastic dynamic coverage approaches include SLAM [74, 28, 21, 90] and information-theoretic methods [49, 50]. A similar filter-based coverage problem is addressed in [49] (for spatially decoupled processes) from an information-theoretic perspective.

2.4.1 Setup and Sensor Model

The grid-based method is used to develop the coverage control problem under probabilistic frameworks. The search domain is discretized into N_{tot} cells. Let $\tilde{c}$ be an arbitrary cell in $\mathscr{D}$, and point $\tilde{q}$ is the centroid of $\tilde{c}$, which is consistent with the definition of $\tilde{q}$ under deterministic Lyapunov-based and aware-based frameworks. Assume that the discretization is fine enough such that at most one object can exist within a cell. The work presented in this section is analogous to the binary Bayes filter and the occupancy grid mapping algorithm [124], which are very popular mapping techniques to deal with observations with sensor uncertainties in robotics.

For the sake of clarity of ideas, first consider the case where there exists a single autonomous sensor-equipped vehicle that performs the search task. This scenario is an extreme case in which the resources available are at a minimum (a single sensor vehicle as opposed to multiple cooperating ones). The extension to MAVs domain search is provided in Section 2.4.5.

A Bernoulli-type sensor model is used, which gives binary outputs: object "present" or "absent". This is a simplified but reasonable sensor model because it abstracts away the complexities in sensor noise, image processing algorithm errors, etc. [12, 13].

Let $X(\tilde{c})$ be the binary state random variable, where $X(\tilde{c}) = 0$ corresponds to object absent and $X(\tilde{c}) = 1$ corresponds to object present. Let the position of object $\mathscr{O}_k$ be $\mathbf{p}_k$ and $\mathscr{P}$ is the set of all object positions (unknown and randomly generated). The number of objects N_o is a Binomial random variable with parameters N_{tot} and $\text{Prob}(\tilde{c} \in \mathscr{P})$, where $\text{Prob}(\tilde{c} \in \mathscr{P})$ is the probability of object presence at cell $\tilde{c}$ (identical for all $\tilde{c}$ and independent). Hence, the probability of k cells in the domain containing an object is

$$\mathrm{Prob}(N_{\mathrm{o}} = k) = \binom{N_{\mathrm{tot}}}{k} \mathrm{Prob}(\tilde{\mathbf{c}} \in \mathscr{P})^{k} (1 - \mathrm{Prob}(\tilde{\mathbf{c}} \in \mathscr{P}))^{N_{\mathrm{tot}} - k},$$

where $k = 1, 2, \cdots, N_{\mathrm{tot}}$. The expectation of N_{o} equals to the number of total cells in $\mathscr{D}$ multiplied by $\mathrm{Prob}(\tilde{\mathbf{c}} \in \mathscr{P})$, that is,

$$E[N_{\mathrm{o}}] = N_{\mathrm{tot}} \mathrm{Prob}(\tilde{\mathbf{c}} \in \mathscr{P}). \tag{2.29}$$

Note that the realization of $X(\tilde{\mathbf{c}})$ depends on the position of the observed cell, that is,

$$X(\tilde{\mathbf{c}}) = \begin{cases} 1 & \tilde{\mathbf{c}} \in \mathscr{P}, \\ 0 & \text{otherwise.} \end{cases}$$

Since $\mathscr{P}$ is unknown and random, $X(\tilde{\mathbf{c}})$ is a random variable with respect to every $\tilde{\mathbf{c}} \in \mathscr{D}$.

Similarly, let $Y(\tilde{\mathbf{c}})$ be the binary observation random variable, where $Y(\tilde{\mathbf{c}}) = 0$ corresponds to the observation indicating object absent and $Y(\tilde{\mathbf{c}}) = 1$ corresponds to the observation indicating object present, respectively. The actual observation is taken according to the probability parameter β of the Bernoulli distribution.

Given a state $X(\tilde{\mathbf{c}}) = j$, the conditional probability mass function f of the Bernoulli observation distribution is given by

$$f_{Y(\tilde{\mathbf{c}})}(Y(\tilde{\mathbf{c}}) = k | X(\tilde{\mathbf{c}}) = j) = \begin{cases} \beta & \text{if } k = j \\ 1 - \beta & \text{if } k \neq j \end{cases}, \quad j, k = 0, 1. \tag{2.30}$$

Because the states $X(\tilde{\mathbf{c}})$ are spatially i.i.d., the observations $Y(\tilde{\mathbf{c}})$ taken at every cell $\tilde{\mathbf{c}}$ within the mission domain $\mathscr{D}$ are spatially i.i.d. and hence the probability distribution for every $\tilde{\mathbf{c}} \in \mathscr{D}$ follows the same structure.

Therefore, the general conditional probability matrix B is given as follows

$$B = \begin{bmatrix} \mathrm{Prob}(Y(\tilde{\mathbf{c}}) = 0 | X(\tilde{\mathbf{c}}) = 0) = \beta & \mathrm{Prob}(Y(\tilde{\mathbf{c}}) = 0 | X(\tilde{\mathbf{c}}) = 1) = 1 - \beta \\ \mathrm{Prob}(Y(\tilde{\mathbf{c}}) = 1 | X(\tilde{\mathbf{c}}) = 0) = 1 - \beta & \mathrm{Prob}(Y(\tilde{\mathbf{c}}) = 1 | X(\tilde{\mathbf{c}}) = 1) = \beta \end{bmatrix}, \tag{2.31}$$

where $\mathrm{Prob}(Y(\tilde{\mathbf{c}}) = i | X(\tilde{\mathbf{c}}) = j)$, $i, j = 0, 1$, describes the probability of measuring $Y(\tilde{\mathbf{c}}) = i$ given state $X(\tilde{\mathbf{c}}) = j$. For the sake of simplicity, it is assumed that the sensor capabilities of making a correct measurement are the same. That is, $\mathrm{Prob}(Y(\tilde{\mathbf{c}}) = 0 | X(\tilde{\mathbf{c}}) = 0) = \mathrm{Prob}(Y(\tilde{\mathbf{c}}) = 1 | X(\tilde{\mathbf{c}}) = 1) = \beta$ as the detection probability of the sensor.

The following two sensor models are assumed in this book. However, note that the specific formulation will not affect the analysis of the subsequent search methods. Both of the sensor models capture the key feature of limited sensory range and will be used interchangeably throughout this book.

Unit Sensory Range

For the sake of illustration clarity, assume that the sensor is only capable of observing one cell at a time. That is, the sensor model assumes a limited unit sensory range. Therefore, β is set as a constant value.

Limited Circular Sensory Domain

To be consistent with the sensor models used in the deterministic frameworks in Sections 2.2 and 2.3, let the detection probability β to be a function of the relative distance between the sensor and the centroid of the observing cell $\tilde{c}$. Similar as the sensor model **SM** proposed in Section 2.2, here a limited-range circular sensor domain is assumed and a fourth order polynomial function of $s = \|q(t) - \tilde{q}\|$ is used within the sensor range r and $b_n = 0.5$ otherwise,

$$\beta(s) = \begin{cases} \frac{M}{r^4}\left(s^2 - r^2\right)^2 + b_n & \text{if } s \leq r \\ b_n & \text{if } s > r \end{cases}, \tag{2.32}$$

where $M + b_n$ gives the peak value of β if the cell $\tilde{c}$ being observed is located at the sensor vehicle's location, which indicates that the sensor's detection probability is highest exactly where it is. The sensing capability decreases with range and becomes 0.5 outside of $\mathcal{W}$, implying that the sensor returns an equal-likely observation of "absence" or "presence" regardless of the truth.

2.4.2 Bayes Updates

Next, Bayes' rule is employed to update the probability of object existence at $\tilde{c}$. Given an observation $Y_t(\tilde{c}) = i$ taken at time step t, Bayes' rule gives, for each $\tilde{c}$, the posterior probability of object existence $(X(\tilde{c}) = j)$ as:

$$P(X(\tilde{c}) = j | Y_t(\tilde{c}) = i; t+1) = \frac{P(Y_t(\tilde{c}) = i | X(\tilde{c}) = j; t)P(X(\tilde{c}) = j; t)}{P(Y_t(\tilde{c}) = i)}, \tag{2.33}$$

where $P(Y_t(\tilde{c}) = i | X(\tilde{c}) = j; t)$ is the probability of the particular observation $Y_t(\tilde{c}) = i$ being taken given state $X(\tilde{c}) = j$, which is given by the β function (2.32), $P(Y_t(\tilde{c}) = i | X(\tilde{c}) = j; t)$ is the prior probability of $X(\tilde{c}) = j$ at t, and $P(Y_t(\tilde{c}) = i)$ gives the total probability of having observation $Y_t(\tilde{c}) = i$ regardless of the actual state.

According to the law of total probability,

$$P(Y_t(\tilde{c}) = i) = P(Y_t(\tilde{c}) = i | X(\tilde{c}) = j; t)P(X = j; t)$$
$$+ P(Y_t(\tilde{c}) = i | X(\tilde{c}) = 1 - j; t)P(X = 1 - j; t), \; i, j = 0, 1. \tag{2.34}$$

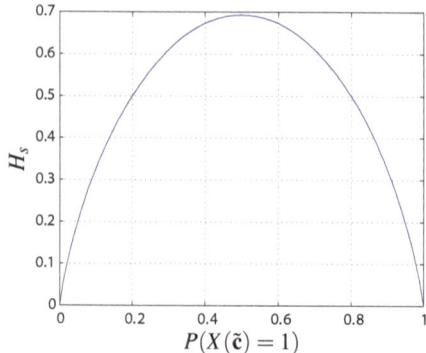

Fig. 2.19 Information entropy function H_s.

Substitute Equation (2.34) into Equation (2.33), the posterior probability of object absent is $P(X(\tilde{c}) = 0|Y_t(\tilde{c}) = i; t + 1)$ and object present is $P(X(\tilde{c}) = 1|Y_t(\tilde{c}) = i; t + 1)$ whenever there is a new observation $Y_t(\tilde{c}) = i$ taken.

2.4.3   Uncertainty Map

Based on the updated probabilities, an information-based approach is used to construct the uncertainty map for every $\tilde{c}$ within the search domain. The uncertainty map will be used to guide the vehicle towards regions with highest search uncertainty in the domain. The information entropy function of a probability distribution is used to evaluate uncertainty. Let $P_{H_s} = \{P(X(\tilde{c}) = 0), P(X(\tilde{c}) = 1)\}$ be the probability distribution of the search process for the two distinct realizations of the state in our case. Define the information entropy at $\tilde{c}$ at time t as:

$$H_s(P_{H_s}, \tilde{c}, t) = -P(X(\tilde{c}) = 0) \ln P(X(\tilde{c}) = 0) - P(X(\tilde{c}) = 1) \ln P(X(\tilde{c}) = 1) \quad (2.35)$$

If $P(X(\tilde{c}) = 0) = 0$, the term $P(X(\tilde{c}) = 0) \ln P(X(\tilde{c}) = 0)$ is set to 0 by convention because there is no uncertainty about object existence or lack thereof. It also follows that $\lim_{P(X(\tilde{c})=0)\to 0} P(X(\tilde{c}) = 0) \ln P(X(\tilde{c}) = 0) = 0$. The same applies for $P(X(\tilde{c}) = 1) \ln P(X(\tilde{c}) = 1)$ when $P(X(\tilde{c}) = 1) = 0$. Figure 2.19 shows the information entropy (2.35) as a function of $P(X(\tilde{c}) = 1)$. Note that $H_s(P_{H_s}, \tilde{c}, t) \geq 0$ and the maximum value attainable by $H_s(P_{H_s}, \tilde{c}, t)$ is $H_{s,max} = 0.6931$ when $P(X(\tilde{c}) = 1) = 0.5$. This implies that the equal-likely case results in the most uncertain information. The information entropy distribution at time step t over the domain forms an uncertainty map at that time instant.

The greater the value of H_s, the bigger the uncertainty is. The desired uncertainty level is $H_s(P_{H_s}, \tilde{c}, t) = 0$ over $\mathcal{D}$. The initial "uncertainty" distribution is assumed to be the maximum value $H_{s,max} \forall \tilde{c} \in \mathcal{D}$ reflecting the fact that at the outset of the search mission there is a poor search certainty level everywhere within the domain.

2.4.4 Bayesian-Based Coverage Control

Now consider a search strategy for the coverage control problem in the Bayesian-based probabilistic framework. In general, the control $\mathbf{u}(t)$ is restricted to a set $\mathscr{U}$. For example, $\mathscr{U}$ could be the set of all controls $\mathbf{u}(t) \in \mathbb{R}^2$ such that $\|\mathbf{u}(t)\| < u_{\max}$, where $u_{\max}$ is the maximum allowable control velocity. First consider a set $\mathscr{Q}_{\mathscr{W}}(t)$. Let

$$\mathscr{Q}_{\mathscr{W}}(t) = \{\tilde{\mathbf{c}} \in \mathscr{W} : \tilde{\mathbf{q}} - \mathbf{q}(t) \in \mathscr{U}\}.$$

In other words, $\mathscr{Q}_{\mathscr{W}}(t)$ is the set of cells within the sensory domain where the vehicle could reach given the restrictions on control.

Consider the following condition, whose utility will become obvious shortly.

Condition C5. $H_s(P_{H_s}, \tilde{\mathbf{c}}, t) \leq H_s^u$, $\forall \tilde{\mathbf{c}} \in \mathscr{Q}_{\mathscr{W}}(t)$, where $H_s^u > 0$ is a preset threshold of some small value.

Following the same structure as the deterministic Lyapunov-based and awareness-based control laws (2.2.1,2.26), the Bayesian-based probabilistic search strategy is given as follows:

$$\mathbf{u}^*(t) = \begin{cases} \bar{\mathbf{u}}(t) & \text{if Condition } \mathbf{C5} \text{ does not hold} \\ \bar{\bar{\mathbf{u}}}(t) & \text{if Condition } \mathbf{C5} \text{ holds} \end{cases}. \qquad (2.36)$$

Let $\tilde{\mathbf{c}}_\star$ be the cell that has the highest search uncertainty within $\mathscr{Q}_{\mathscr{W}}(t)$, that is,

$$\tilde{\mathbf{c}}_\star(t+1) = \operatorname{argmax}_{\tilde{\mathbf{c}} \in \mathscr{Q}_{\mathscr{W}}(t)} H_s(P_{H_s}, \tilde{\mathbf{c}}, t). \qquad (2.37)$$

The nominal control law is then set to be

$$\bar{\mathbf{u}}(t) = \tilde{\mathbf{q}}_\star(t+1) - \mathbf{q}(t) \in \mathscr{U},$$

where $\tilde{\mathbf{q}}_\star$ is the centroid of cell $\tilde{\mathbf{c}}_\star$.

If Condition **C5** holds, then the perturbation controller $\bar{\bar{\mathbf{u}}}(t)$ is used, and $\tilde{\mathbf{q}}^*$ is chosen as the centroid of $\tilde{\mathbf{c}}^* \in \mathscr{Q}_{\mathscr{D}}(t) = \{\tilde{\mathbf{c}} \in \mathscr{D} : \tilde{\mathbf{q}} - \mathbf{q}(t) \in \mathscr{U}\}$ such that $H_s(P_{H_s}, \tilde{\mathbf{c}}^*, t) > H_s^u$.

The choice of $\tilde{\mathbf{c}}^*$ by the vehicle can be made several ways. Here provides one example of many possible perturbation control approaches, which is consistent with the scheme presented in Section 2.2.2. This maneuver seeks the minimum distance for redeployment, and hence is efficient energy-wise than other possibilities. Let

$$\mathscr{D}_e(t) := \{\tilde{\mathbf{c}} \in \mathscr{Q}_{\mathscr{D}}(t) : H_s(P_{H_s}, \tilde{\mathbf{c}}, t) > H_s^u\},$$

which is a set of all $\tilde{\mathbf{c}}$ for which $H_s(P_{H_s}, \tilde{\mathbf{c}}, t)$ is larger than the preset value H_s^u. Let $\widetilde{\mathscr{D}}_e(t)$ be the set of cells in $\mathscr{D}_e(t)$ that minimize the distance between the position vector of vehicle $\mathscr{V}$, $\mathbf{q}$, and the set $\mathscr{D}_e(t)$:

$$\widetilde{\mathscr{D}}_e(t) = \left\{ \tilde{\mathbf{c}}^* \in \mathscr{D}_e(t) : \tilde{\mathbf{c}}^* = \mathrm{argmin}_{\tilde{\mathbf{c}} \in \mathscr{D}_e(t)} \|\tilde{\mathbf{q}} - \mathbf{q}(t)\| \right\}.$$

The control law (2.36) guarantees that the uncertainty function $H_s(P_{H_s}, \tilde{\mathbf{c}}, t_s)$ for all $\tilde{\mathbf{c}} \in \mathscr{D}$ is below H_s^u at some time t_s. A formal proof will be given as part of the Bayesian-based decision-making strategy in Section 4.3 in Chapter 4.

Remark 2.4.1. *Note that according to Equation (2.37), $\tilde{\mathbf{c}}_*(t+1)$ might be a set of cells holding the same maximum search uncertainty value. If there are multiple such cells, then one can define a rule that picks the "best" one according to some metric (e.g., the cell with its centroid that is closest to the vehicle's current position). Currently, assume there is only one such cell for the sake of simplicity.* •

Remark 2.4.2. *The reasons that the choice of $\tilde{\mathbf{c}}_*$ is restricted to $\mathscr{W}$ (as opposed to $\mathscr{D}$) in the definition of $\mathscr{Q}_{\mathscr{W}}(t)$ (causing $\bar{\mathbf{u}}$ to become a local controller) are as follows:*

1. *Using $\mathscr{W}$ instead of $\mathscr{D}$ avoids unnecessary extra computational burden during the search for $\tilde{\mathbf{c}}_*$ by using a smaller space and, hence, is more computationally efficient. It is especially true in the case of large-scale domains, where much of the domain $\mathscr{D}$ is unreachable from where the vehicle is because of the restriction on $\mathbf{u}$ to be in the control set $\mathscr{U}$.*
2. *Although in this book it is assumed that the vehicle has full knowledge of the domain $\mathscr{D}$ and the search uncertainty function $H_s(P_{H_s}, \tilde{\mathbf{c}}, t)$ for all $\tilde{\mathbf{c}} \in \mathscr{D}$, $\mathscr{D}$ may not be known in real time. In this case, all the information the vehicle could obtain is within its limited sensory domain $\mathscr{W}$.* •

Remark 2.4.3. *Having $\mathscr{U}$ arbitrary (i.e., such that $\mathscr{Q}_{\mathscr{D}}(t)$ may not be equal to $\mathscr{D}$), our algorithm may get stuck in regions where $H_s < H_s^u$ and no control can take outside this region and no overall coverage can be guaranteed. This is a shortcoming of the current proposed control strategy but as long as there is no global centralized computer that sees the entire $\mathscr{D}$, there is very little any control policy will ever be able to do.* ◢

2.4.5 Extension to MAVs with Intermittent Information Sharing

In this section, the Bayesian-based domain search strategies are extended to distributed MAVs with intermittent information sharing. Multi-sensor fusion based on observations from neighboring vehicles is implemented via binary Bayes filter. It will be proved that, under appropriate sensor models, the belief of whether objects exist or not will converge to the true state. Different motion control schemes are numerically tested to illustrate the effectiveness of the proposed strategy.

In order to reduce the uncertainty due to sensor errors, or equivalently, to maximize the probability of finding an object of interest, all the available observations a vehicle has access to (i.e., taken by the vehicle itself and its neighboring vehicles) should be fused together and utilized as a combined observation sequence. It will be proved that given sensors with a detection probability greater than 0.5, the search uncertainty will converge to a small neighborhood of zero, i.e., all unknown objects of interest are found with 100% confidence level. This is a nontrivial problem given limited theoretical results existing in the literature and its significance for effective sensor management, especially when the sensing and communication resources are limited.

2.4.5.1 Bayes Updates for MAVs with Intermittent Communications

Let the detection probability of each vehicle sensor $\mathcal{V}_i$ be denoted as β^i. Clearly, $\beta^i \in [0,1]$. In this section, the binary Bayes filter is employed to update the probability of object presence at $\tilde{c}$ of vehicle $\mathcal{V}_i$ based on all the observations available at the current time step and the prior probability. Define $\bar{Y}_t^i(\tilde{c}) = \{\mathcal{V}_j \in \mathcal{G}_i(t) : Y_{j,t}(\tilde{c})\}$ as the observation sequence taken by all the vehicles in vehicle $\mathcal{V}_i$'s neighborhood $\mathcal{G}_i(t)$ at time t. Given $\bar{Y}_t^i(\tilde{c})$, Bayes' rule gives, for each vehicle $\mathcal{V}_i$,

$$P_i(X(\tilde{c}) = 1|\bar{Y}_t^i(\tilde{c}); t+1) = \eta_i P_i(\bar{Y}_t^i(\tilde{c})|X(\tilde{c}) = 1) P_i(X(\tilde{c}) = 1; t),$$

where $P_i(X(\tilde{c}) = 1|\bar{Y}_t^i(\tilde{c}); t+1)$ is the posterior probability of object presence at cell $\tilde{c}$ updated by vehicle $\mathcal{V}_i$ after the observation sequence has been taken at time step t. The quantity $P_i(\bar{Y}_t^i(\tilde{c})|X(\tilde{c}) = 1)$ is the probability of the particular observation sequence $\bar{Y}_t^i$ being taken given that the actual state at cell $\tilde{c}$ is object present. Because the observations taken by different vehicles are i.i.d., it follows that $P_i(\bar{Y}_t^i(\tilde{c})|X(\tilde{c}) = 1) = \Pi_{j \in \mathcal{G}_i(t)} \text{Prob}(Y_{j,t}(\tilde{c})|X(\tilde{c}) = 1)$, where $\text{Prob}(Y_{j,t}(\tilde{c})|X(\tilde{c}) = 1)$ is given by the conditional probability matrix (2.31) and the Bernoulli observation distribution (2.30). The quantity $P_i(X(\tilde{c}) = 1; t)$ is the prior probability of object presence at time t, and η_i serves as a normalizing function which ensures that the posterior probabilities $P_i(X(\tilde{c}) = j|\bar{Y}_t^i(\tilde{c}); t+1)$, $j = 0, 1$ sum to one.

According to the law of total probability, the posterior probability of object presence at $\tilde{c}$ updated according to all the observations available to vehicle $\mathcal{V}_i$ is given by the following equation, where $y_{j,t}(\tilde{c})$ is the dummy variable for the random variable $Y_{j,t}(\tilde{c})$.

$$P_i(X(\tilde{c}) = 1|\bar{Y}_t^i(\tilde{c}); t+1)$$
$$= \frac{P_i(X(\tilde{c}) = 1; t)}{P_i(X(\tilde{c}) = 1; t) + \Pi_{j \in \mathcal{G}_i(t)} \left(\left(\frac{1}{\beta^j} - 1\right)^{2y_{j,t}(\tilde{c})-1} \right) (1 - P_i(X(\tilde{c}) = 1; t))}. \quad (2.38)$$

Note that the probability of object absence is given by

$$P_i(X(\tilde{c}) = 0|\bar{Y}_t^i(\tilde{c}); t+1) = 1 - P_i(X(\tilde{c}) = 1|\bar{Y}_t^i(\tilde{c}); t+1).$$

2.4.5.2   Convergence Analysis

In this section, the conditions for convergence of the sequence $\{P_i(X(\tilde{c}) = 1|\bar{Y}_t^i(\tilde{c});t+1)\}$ is discussed when β^i, $i = 1,2,\cdots,N$ is a deterministic parameter within $[0,1]$.

For the sake of simplicity, denote $P_i(X(\tilde{c}) = 1|\bar{Y}_t^i(\tilde{c});t+1)$ as P_{t+1}, $P_i(X(\tilde{c}) = 1;t)$ as P_t, and $\Pi_{j\in\mathscr{G}_i(t)}\left((\frac{1}{\beta^j}-1)^{2y_{j,t}(\tilde{c})-1}\right)$ as S_t, Equation (2.38) then simplifies to the following non-autonomous nonlinear discrete-time system

$$P_{t+1} = \frac{P_t}{P_t + S_t(1-P_t)}. \tag{2.39}$$

Note that S_t is a random variable dependent on the observation sequence $\bar{Y}_t^i(\tilde{c})$. Let $|\mathscr{G}_i(t)|$ be the cardinality of the set $\mathscr{G}_i(t)$, i.e., the number of neighboring vehicles of vehicle $\mathscr{V}_i$ and itself, then the binary observation sequence $\bar{Y}_t^i(\tilde{c})$ has $2^{|\mathscr{G}_i(t)|}$ possible combinations at each time step t for cell $\tilde{c}$. Let $s_t^1, s_t^2, \cdots, s_t^{2^{|\mathscr{G}_i(t)|}}$ be the realizations of S_t corresponding to each of the $2^{|\mathscr{G}_i(t)|}$ different observation sequences. The probability of having each particular observation sequence $\bar{Y}_t^i(\tilde{c}) = \{Y_{j,t}(\tilde{c}) = y_{j,t}(\tilde{c}), \mathscr{V}_j \in \mathscr{G}_i(t)\}$ given object present is: $\Pi_{j\in\mathscr{G}_i(t)}(\beta^j)^{y_{j,t}(\tilde{c})}(1-\beta^j)^{(1-y_{j,t}(\tilde{c}))}$.

Consider the following conditional expectation

$$
\begin{aligned}
E[1-P_{t+1}|P_t] &= E[\frac{S_t(1-P_t)}{P_t+S_t(1-P_t)}|P_t = p_t] \\
&= E[\frac{S_t(1-p_t)}{p_t+S_t(1-p_t)}|P_t = p_t] \\
&= \sum_{m=1}^{2^{|\mathscr{G}_i(t)|}} \frac{s_t^m(1-p_t)}{p_t+s_t^m(1-p_t)}\mathrm{Prob}(S_t = s_t^m|P_t = p_t), \tag{2.40}
\end{aligned}
$$

where p_t is the dummy variable for P_t. The substitution law and the law of total probability are used in the above derivation. Because the observation sequence taken at each time step is a property of the sensors, and is not affected by the probability of object presence at the previous time step, S_t is independent of P_t. Therefore, Equation (2.40) can be reduced to

$$E[1-P_{t+1}|P_t] = \sum_{m=1}^{2^{|\mathscr{G}_i(t)|}} \frac{s_t^m(1-p_t)}{p_t+s_t^m(1-p_t)}\mathrm{Prob}(S_t = s_t^m). \tag{2.41}$$

Investigate the value of s_t^m and the corresponding $\mathrm{Prob}(S_t = s_t^m)$ from $m = 1$ to $2^{|\mathscr{G}_i(t)|}$.

- $m = 1$ corresponds to the observation sequence $\{1,1,\cdots,1\}$, it follows that $s_t^1 = \Pi_{j\in\mathscr{G}_i(t)}(\frac{1}{\beta^j}-1)$ and $\mathrm{Prob}(S_t = s_t^1) = \Pi_{j\in\mathscr{G}_i(t)}\beta^j$
- $m = k+1$, $k = 1,\cdots,|\mathscr{G}_i(t)|$ correspond to the observation sequence where only the k_{th} vehicle in vehicle $\mathscr{V}_i$'s neighborhood observes a 0. Define C_k^n as the binomial coefficients, i.e., the number of combinations that one can choose k objects

from a set of size n. Because there are $C_1^{|\mathscr{G}_i(t)|}$ such observation sequences with different orders out of the totally $2^{|\mathscr{G}_i(t)|}$ combinations, the value of k is in the set $[1, |\mathscr{G}_i(t)|]$. Hence, it follows that $s_t^{k+1} = \left(\Pi_{j \in \mathscr{G}_i(t),\ j \neq k} (\frac{1}{\beta^j} - 1) \right) \left(\frac{\beta^k}{1-\beta^k} \right)$ and $\text{Prob}(S_t = s_t^{k+1}) = \left(\Pi_{j \in \mathscr{G}_i(t),\ j \neq k} \beta^j \right) (1 - \beta^k)$

- $m = k+1+|\mathscr{G}_i(t)|,\ k = 1, \cdots, C_2^{|\mathscr{G}_i(t)|}$ correspond to the the observation sequences where two of the vehicles, e.g., the q_{th} and r_{th} vehicle, observe a 0. Because there are $C_2^{|\mathscr{G}_i(t)|}$ such observation sequences in this case, k is within $[1, C_2^{|\mathscr{G}_i(t)|}]$. Therefore, it follows that

$$s_t^{k+1+|\mathscr{G}_i(t)|} = \left(\Pi_{j \in \mathscr{G}_i(t),\ j \neq q,r} (\frac{1}{\beta^j} - 1) \right) \left(\frac{\beta^q}{1-\beta^q} \right) \left(\frac{\beta^r}{1-\beta^r} \right)$$

and
$\text{Prob}(S_t = s_t^{k+1+|\mathscr{G}_i(t)|}) = \left(\Pi_{j \in \mathscr{G}_i(t),\ j \neq q,r} \beta^j \right) (1 - \beta^q)(1 - \beta^r)$
- And so on for other values of m
- $m = 2^{|\mathscr{G}_i(t)|}$ correspond to the observation sequence $\{0, 0, \cdots, 0\}$, it follows that $s_t^m = \Pi_{j \in \mathscr{G}_i(t)} \left(\frac{\beta^j}{1-\beta^j} \right)$ and $\text{Prob}(S_t = s_t^m) = \Pi_{j \in \mathscr{G}_i(t)} (1 - \beta^j)$

Suppose $p_t = 1 - \varepsilon$, where $\varepsilon \in [0, \frac{1}{2})$ is some constant, Equation (2.41) can be rewritten as the follows if not all sensing parameters $\beta^j = 1$, and $E[1 - P_{t+1}|P_t] = 0$ when all $\beta^j = 1,\ j \in \mathscr{G}_i(t)$.

$$E[1 - P_{t+1}|P_t] = \left(\frac{\Pi_{j \in \mathscr{G}_i(t)}(1 - \beta^j)}{1 - \varepsilon + \Pi_{j \in \mathscr{G}_i(t)}(\frac{1}{\beta^j} - 1)\varepsilon} + \sum_{k=1}^{|\mathscr{G}_i(t)|} \frac{\Pi_{j \in \mathscr{G}_i(t)}(1 - \beta^j)}{(\frac{1}{\beta^k} - 1)(1 - \varepsilon) + \Pi_{j \in \mathscr{G}_i(t),\ j \neq k}(\frac{1}{\beta^j} - 1)\varepsilon} + \ldots + \right.$$

$$\left. \sum_{k=1}^{C_2^{|\mathscr{G}_i(t)|}} \frac{\Pi_{j \in \mathscr{G}_i(t)}(1 - \beta^j)}{(\frac{1}{\beta^q} - 1)(\frac{1}{\beta^r} - 1)(1 - \varepsilon) + \Pi_{j \in \mathscr{G}_i(t),\ j \neq q \cdot r}(\frac{1}{\beta^j} - 1)\varepsilon} + \ldots + \frac{\Pi_{j \in \mathscr{G}_i(t)}(1 - \beta^j)}{\Pi_{j \in \mathscr{G}_i(t)}(\frac{1}{\beta^j} - 1)(1 - \varepsilon) + \varepsilon} \right) \varepsilon.$$

$$(2.42)$$

Consider the following condition:

Sensing Condition 1: $\beta^i \in (\frac{1}{2}, 1],\ i = 1, 2, \cdots, N$.

This condition requires that all vehicle sensors are more likely to take correct measurements.

Under Sensing Condition 1, it follows that $\Pi_{j \in \mathscr{G}_i(t)}(\frac{1}{\beta^j} - 1) \in [0, 1)$. Now assume that ε is a small number in the neighborhood of zero, given $\Pi_{j \in \mathscr{G}_i(t)}(\frac{1}{\beta^j} - 1)\varepsilon$ is also a small number close to zero, hence, the conditional expectation given by Equation (2.42) can be approximated as

$$E[1 - P_{t+1}|P_t] \approx \left(\Pi_{j \in \mathscr{G}_i(t)}(1 - \beta^j) + \sum_{k=1}^{|\mathscr{G}_i(t)|} \Pi_{j \in \mathscr{G}_i(t),\ \neq k}(1 - \beta^j)\beta^k \right.$$

$$\left. + \ldots + \sum_{k=1}^{C_2^{|\mathscr{G}_i(t)|}} \Pi_{j \in \mathscr{G}_i(t),\ \neq q,r}(1 - \beta^j)\beta^q \beta^r + \ldots + \Pi_{j \in \mathscr{G}_i(t)}\beta^j \right) \varepsilon. \quad (2.43)$$

Observe the expression within the bracket in Equation (2.43), it gives the total probability of all possible observation sequences taken by the vehicles in set $\mathscr{G}_i(t)$ given that there is an object within $\tilde{c}$, and is therefore equal to 1. If $\beta^j = \beta$, $\forall \mathscr{V}_j \in \mathscr{G}_i(t)$, the expression gives the total probability of a binomial distribution with parameter β and $|\mathscr{G}_i(t)|$.

Hence, the conditional probability $E[1 - P_{t+1}|P_t = 1 - \varepsilon] \approx \varepsilon$ and the following lemma holds.

Lemma 2.4.1. *Under Sensing Condition 1, if an object is present, given that the prior probability of object presence $P_i(X(\tilde{c}) = 1;t)$ of vehicle $\mathscr{V}_i$ is within a small neighborhood of radius ε from 1 at time step t, then the conditional expectation of the posterior probability $P_i(X(\tilde{c}) = 1|\bar{Y}_t^i(\tilde{c});t+1)$ will remain in this neighborhood at the next time step. If all the sensors are "perfect" with zero observation error probability, i.e., $\beta^j = \beta = 1$, then the conditional expectation of $P_i(X(\tilde{c}) = 1|\bar{Y}_t^i(\tilde{c});t+1)$ is 1.*

Following a similar derivation as above, a lemma holds for the posterior probability of object absence $P_i(X(\tilde{c}) = 0|\bar{Y}_t^i(\tilde{c});t+1)$ given there is no object at cell $\tilde{c}$. Note that in this case, if abusing notation and still denoting $P_i(X(\tilde{c}) = 0|\bar{Y}_t^i(\tilde{c});t+1)$ as P_{t+1} and $P_i(X(\tilde{c}) = 0;t)$ as P_t, it follows that $S_t = \Pi_{j \in \mathscr{G}_i(t)} \left(\frac{\beta^j}{1 - \beta^j} \right)^{2y_{j,t}(\tilde{c})-1}$ and the probability of having each particular observation sequence given object absent is $\Pi_{j \in \mathscr{G}_i(t)} (1 - \beta^j)^{y_{j,t}(\tilde{c})} (\beta^j)^{(1 - y_{j,t}(\tilde{c}))}$.

To summarize the above results, the following theorem holds.

Theorem 2.4.1. *For $\beta^i \in (\frac{1}{2}, 1]$, $i = 1, 2, \cdots, N$, if there is an object absent (respectively, present), given that $P_i(X(\tilde{c}) = 0;t)$ (respectively, $P_i(X(\tilde{c}) = 1;t)$) is within a small neighborhood of 1 at time step t, the conditional expectation of $P_i(X(\tilde{c}) = 0|\bar{Y}_t^i(\tilde{c});t+1)$ (respectively, $P_i(X(\tilde{c}) = 1|\bar{Y}_t^i(\tilde{c});t+1)$) will remain in this neighborhood at the next time step. If $\beta^i = \beta = 1$, then the conditional expectation is 1.*

This theorem gives a weak result because it implies that only if the initial prior probability is close to the true state, given "good" sensors with detection probabilities greater than 0.5, the belief of whether objects exist or not will remain near the true state. Next, a stronger result is derived for the case of homogeneous sensor properties across the network.

Next, consider the following condition.

Sensing Condition 2: $\beta^i = \beta \in (\frac{1}{2}, 1]$, $i = 1, 2, \cdots, N$.

This condition implies that all the vehicles have identical sensors with the same detection probability $\beta \in (\frac{1}{2}, 1]$.

Under Sensing Condition 2, the term within the bracket in Equation (2.42) is equivalent to the following expression:

$$g(\beta, \varepsilon, |\mathscr{G}_i(t)|) = \sum_{k=0}^{|\mathscr{G}_i(t)|} \frac{C_k^{|\mathscr{G}_i(t)|} (1 - \beta)^{|\mathscr{G}_i(t)|}}{(\frac{1}{\beta} - 1)^k (1 - \varepsilon) + (\frac{1}{\beta} - 1)^{|\mathscr{G}_i(t)| - k} \varepsilon}, \quad \beta \neq 1. \quad (2.44)$$

Lemma 2.4.2. *The function $g(\beta, \varepsilon, |\mathcal{G}_i(t)|)$ is less than 1 for $\beta \in (\frac{1}{2}, 1)$, $\varepsilon \in (0, \frac{1}{2})$, and $|\mathcal{G}_i(t)| \geq 1$. Moreover, $g(\beta, \varepsilon, |\mathcal{G}_i(t)|) = 1$ for $\varepsilon = 0$, $\beta \in (\frac{1}{2}, 1)$, and $|\mathcal{G}_i(t)| \geq 1$.*

Before providing a rigorous proof, the results shown in Figure 2.20 confirm the above lemma. Figure 2.20 shows $g(\beta, \varepsilon, |\mathcal{G}_i(t)|)$ as a function of $\varepsilon \in [0, \frac{1}{2})$ and $\beta \in (\frac{1}{2}, 1)$ for (a) $|\mathcal{G}_i(t)| = 1$, (b) $|\mathcal{G}_i(t)| = 20$, (c) $|\mathcal{G}_i(t)| = 50$, and (d) $|\mathcal{G}_i(t)| = 100$. It can be seen that g is less than or equal to 1 for $\varepsilon \in [0, \frac{1}{2})$ and $\beta \in (\frac{1}{2}, 1)$.

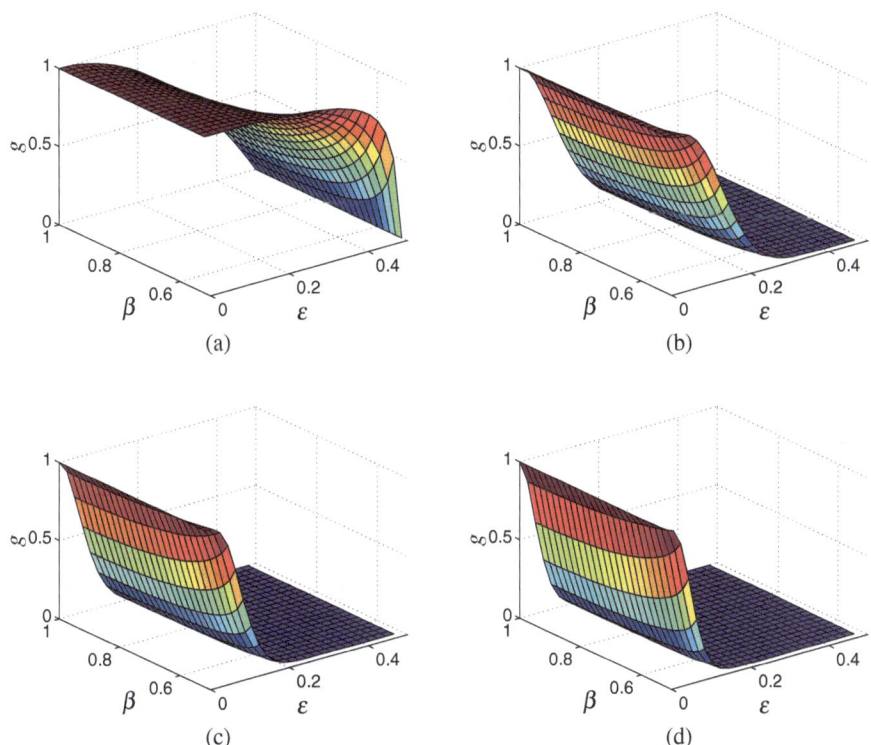

Fig. 2.20 $g(\beta, \varepsilon, |\mathcal{G}_i(t)|)$ as a function of ε and β.

The following gives the proof for Lemma 2.4.2.

Proof. For brevity, let $n = |\mathcal{G}_i(t)|$. Proving that $g(\beta, \varepsilon, |\mathcal{G}_i(t)|)$ is less than 1 is equivalent to prove that

$$\sum_{k=0}^{n} \frac{C_k^n (1-\beta)^n}{(\frac{1}{\beta}-1)^k(1-\varepsilon)+(\frac{1}{\beta}-1)^{n-k}\varepsilon} < \sum_{k=0}^{n} \frac{1}{n+1}, \text{ or,}$$

$$\sum_{k=0}^{n} \frac{(n+1)C_k^n(1-\beta)^n - \left[(\frac{1}{\beta}-1)^k(1-\varepsilon)+(\frac{1}{\beta}-1)^{n-k}\varepsilon\right]}{\left[(\frac{1}{\beta}-1)^k(1-\varepsilon)+(\frac{1}{\beta}-1)^{n-k}\varepsilon\right](n+1)} < 0.$$

Because $\beta \in (\frac{1}{2}, 1)$, or $(\frac{1}{\beta}-1) \in (0,1)$, it follows that

$$\sum_{k=0}^{n} \frac{(n+1)C_k^n(1-\beta)^n - \left[(\frac{1}{\beta}-1)^k(1-\varepsilon)+(\frac{1}{\beta}-1)^{n-k}\varepsilon\right]}{\left[(\frac{1}{\beta}-1)^k(1-\varepsilon)+(\frac{1}{\beta}-1)^{n-k}\varepsilon\right](n+1)}$$

$$< \sum_{k=0}^{n} \frac{(n+1)C_k^n(1-\beta)^n - \left[(\frac{1}{\beta}-1)^k(1-\varepsilon)+(\frac{1}{\beta}-1)^{n-k}\varepsilon\right]}{\left[(\frac{1}{\beta}-1)^n(1-\varepsilon)+(\frac{1}{\beta}-1)^n\varepsilon\right](n+1)}.$$

Since $\left[(\frac{1}{\beta}-1)^n(1-\varepsilon)+(\frac{1}{\beta}-1)^n\varepsilon\right](n+1) > 0$ for all $k = 0, \cdots, n$, if

$$\sum_{k=0}^{n} (n+1)C_k^n(1-\beta)^n - \left[(\frac{1}{\beta}-1)^k(1-\varepsilon)+(\frac{1}{\beta}-1)^{n-k}\varepsilon\right] < 0,$$

then $g(\beta, \varepsilon, n)$ is less than 1. Note that

$$\sum_{k=0}^{n} (n+1)C_k^n(1-\beta)^n = (n+1)(2-2\beta)^n,$$

and

$$\sum_{k=0}^{n} (\frac{1}{\beta}-1)^k = \sum_{k=0}^{n} (\frac{1}{\beta}-1)^{n-k} = \frac{1-(\frac{1}{\beta}-1)^{n+1}}{2-\frac{1}{\beta}},$$

Therefore, to prove the lemma, it only needs to prove that

$$(n+1)(2-2\beta)^n - \frac{1-(\frac{1}{\beta}-1)^{n+1}}{2-\frac{1}{\beta}} < 0. \tag{2.45}$$

Next, the principle of mathematical induction is used to prove the inequality in Equation (2.45).

When $n = 1$, the left hand side of Equation (2.45) is given by $\frac{-(2\beta-1)^2}{\beta}$ and is hence less than 0.

Assume that for $n = m$,

$$(m+1)(2-2\beta)^m - \frac{1-(\frac{1}{\beta}-1)^{m+1}}{2-\frac{1}{\beta}} < 0,$$

therefore, when $n = m + 1$, the left hand side of Equation (2.45) is given by

$$(m+2)(2-2\beta)^{m+1} - \frac{1-(\frac{1}{\beta}-1)^{m+2}}{2-\frac{1}{\beta}} < (m+2)(2-2\beta)\frac{1-(\frac{1}{\beta}-1)^{m+1}}{(m+1)(2-\frac{1}{\beta})} - \frac{1-(\frac{1}{\beta}-1)^{m+2}}{2-\frac{1}{\beta}} \quad (2.46)$$

Skipping all the detailed derivations, it follows that the right hand side of Equation (2.46) is equal to the following expression,

$$\frac{(1-2\beta)m + (3-4\beta) + (\frac{1}{\beta}-1)^{m+2}[(1-2\beta)m + (1-4\beta)]}{(m+1)(2-\frac{1}{\beta})},$$

and it can be shown that the numerator is always less than 0 and the denominator is always larger than 0 for $\beta \in (\frac{1}{2}, 1)$ and $m \geq 1$.

To see why this is true, first when $m = 1$ and $\beta \in (\frac{1}{2}, 1)$, the numerator equals to the following expression

$$(4-6\beta) + \left(\frac{1}{\beta}-1\right)^3 (2-6\beta) < 0.$$

Next, take derivative of the numerator with respect to m, which gives

$$(1-2\beta) + (m+2)\left(\frac{1}{\beta}-1\right)^{m+1}[(1-2\beta)m + (1-4\beta)] + \left(\frac{1}{\beta}-1\right)^{m+2}(1-2\beta) < 0.$$

Therefore, the numerator is a monotonically decreasing function for $m \geq 1$ with a negative value at $m = 1$.

When $\varepsilon = 0$, $g(\beta, \varepsilon, n)$ reduces to

$$\sum_{k=0}^{n} \frac{C_k^n(1-\beta)^n}{(\frac{1}{\beta}-1)^k} = \sum_{k=0}^{n} C_k^n \beta^k (1-\beta)^{n-k} = 1.$$

This completes the proof. □

Therefore, from Lemma 2.4.2, the expectation $E[1 - P_{t+1}|P_t = 1 - \varepsilon]$ is always less than ε. Hence, the following lemma holds.

Lemma 2.4.3. *Under Sensing Condition 2, if there is an object present, given that the prior probability of object presence $P_i(X(\tilde{c}) = 1; t)$ is within a neighborhood of one with radius $\varepsilon \in [0, \frac{1}{2})$, then the conditional expectation of the posterior probability $P_i(X(\tilde{c}) = 1|\bar{Y}_t^i(\tilde{c}); t+1)$ converges to 1.*

Same lemma follows for the update sequence $E[P_i(X(\tilde{c}) = 0|\bar{Y}_t^i(\tilde{c}); t+1)]$. Therefore, the following theorem holds.

Theorem 2.4.2. *For $\beta^i = \beta \in (\frac{1}{2}, 1]$, $i = 1, 2, \cdots, N$, if an object is present (respectively, absent), then $E[P_i(X(\tilde{c}) = 1|\bar{Y}_t^i(\tilde{c}); t+1)]$ converges to 1 (respectively, 0).*

2.4.5.3   Uncertainty and Coverage Metric

From Theorem 2.4.1, it is known that given the true state, the expected posterior probability of object presence/absence $\forall \tilde{c} \in \mathcal{D}$ will be bounded within a small neighborhood of 1 with radius ε if the priors are given by $1 - \varepsilon$. This corresponds to an upper bound on the search uncertainty level $H_{i,s}^u = -\varepsilon \ln \varepsilon - (1 - \varepsilon) \ln(1 - \varepsilon)$. Here, the information entropy function $H_{i,s}$ follows the same form as Equation (2.35) and the subscript i is used to indicate that this is the uncertainty level attained by vehicle $\mathcal{V}_i$. Moreover, from Theorem 2.4.2, it is guaranteed that the expected posterior probability converges to 1, which is equivalent to $H_{i,s} \to 0$, $\forall \tilde{c} \in \mathcal{D}$.

Now, define the coverage metric to evaluate the progress of the search task. Associate each vehicle $\mathcal{V}_i$ with the following search cost function:

$$\mathcal{J}_i(t) = \frac{\sum_{\tilde{c} \in \mathcal{D}} H_{i,s}(P_{i,H_s}, \tilde{c}, t)}{H_{s,\max} A_{\mathcal{D}}}. \tag{2.47}$$

The cost $\mathcal{J}_i(t)$ is proportional to the sum of search uncertainty over $\mathcal{D}$. $\mathcal{J}_i(t)$ is normalized by dividing the sum over all cells by the area of the domain $A_{\mathcal{D}}$ multiplied by $H_{s,\max}$. According to this definition, it follows that $0 \leq \mathcal{J}_i(t) \leq 1$. Initially, $\mathcal{J}_i(t=0) = \frac{H_{i,s}(P_{i,H_s}, \tilde{c}, t)}{H_{s,\max}} \leq 1$. If $H_{i,s}(P_{i,H_s}, \tilde{c}, t_s) = 0$ at some $t = t_s$ for all $\tilde{c} \in \mathcal{D}$, then $\mathcal{J}_i(t_s) = 0$ and the entire domain has been satisfactorily covered and it is 100% certainty that there are no more objects yet to be found.

2.4.5.4   Vehicle Motion Control Scheme

General Motion Control Scheme

According to the search metric (2.47), the upper bound on the uncertainty level $H_{i,s}^u$ results in $\mathcal{J}_i^u(t_f) = \frac{H_{i,s}^u}{H_{s,\max}} = \delta \geq 0$ at some time $t_f > 0$. This is equivalent to say that the attained accuracy of the domain search task is $1 - \delta$. Furthermore, 100% certainty can be obtained if Sensing Condition 2 is satisfied according to Theorem 2.4.2. Therefore, under any vehicle motion control scheme that covers all the cells within the entire mission domain $\mathcal{D}$, the cost function $\mathcal{J}_i \to \delta$, i.e., all the objects of interest will be guaranteed to be found with desired uncertainty. This section seeks vehicle motion control strategies that take advantage of the uncertainty map and perform the search mission efficiently. Two different vehicle motion control schemes that utilize the uncertainty map will be presented, and their performance is compared in simulations. The limited-range circular sensor model is used to model β^i in these control schemes. This sensor model guarantees the realization of Sensing Condition 1. To satisfy Sensing Condition 2, one may assume an identical value $\beta > 0.5$ within $\mathcal{W}_i$ and 0.5 outside of it for all the vehicles.

Memoryless Motion Control Scheme

In this section, first consider a motion control scheme that guides the vehicles based on only the uncertainty map at current time step, that is, the control scheme is memoryless. For the sake of simplicity, assume that there is no speed limit on the vehicles, i.e., a vehicle is able to move to any cell within $\mathscr{D}$ from its current location.

Consider the set

$$\mathscr{Q}_H^i(t) = \{\tilde{c} \in \mathscr{D} : \mathrm{argmax}_{\tilde{c}} H_{i,s}(P_{i,H_s}, \tilde{c}, t_s)\}, \tag{2.48}$$

which is the set of cells with highest search uncertainty level $H_{i,s}$ of vehicle $\mathscr{V}_i$ within $\mathscr{D}$ at time t. Next, let $\tilde{q}_c^i(t)$ be the centroid of the cell that vehicle $\mathscr{V}_i$ is currently located at and define the subset $\mathscr{Q}_d^i(t) \subseteq \mathscr{Q}_H^i(t)$ as

$$\mathscr{Q}_d^i(t) = \{\tilde{c} \in \mathscr{Q}_H^i(t) : \mathrm{argmin}_{\tilde{c}} \|\tilde{q}_c^i(t) - \tilde{q}\|\}, \tag{2.49}$$

where $\tilde{q}$ is the centroid of $\tilde{c}$. The set $\mathscr{Q}_d^i(t)$ contains the cells which have both the shortest distance from the current cell and the highest uncertainty.

At every time step, a vehicle $\mathscr{V}_i$ takes observations at all the cells within its sensory range. In general, $\beta^i \neq \beta^j$, $j \in \mathscr{G}_i(t)$, if $\mathscr{V}_i$ and its neighbor $\mathscr{V}_j$ have same distance to the centroid of a certain cell $\tilde{c}$, it follows that $\beta^i = \beta^j$, $i \neq j$. The posterior probabilities at these cells are updated according to Equation (2.38) based on all the fused observations. The uncertainty map is then updated. At the next time step, the vehicle will choose the next cell to go to from $\mathscr{Q}_d(t)$ based on the updated uncertainty map. Note that $\mathscr{Q}_d(t)$ may have more than one cell. Let N_{Hd} be the number of cells in $\mathscr{Q}_d(t)$, the sensor will randomly pick a cell from $\mathscr{Q}_d(t)$ with probability $\frac{1}{N_{\mathrm{Hd}}}$. This process is repeated until H_i is within a small neighborhood of zero with radius ε for every cell $\tilde{c} \in \mathscr{D}$, which is equivalent to finding all the unknown objects with a desired certainty level.

Motion Control Scheme with Memory

This section develops a motion control scheme that takes into account both the current probability information, uncertainty map and the sensing history. First consider the following condition:

Condition C6: $H_{i,s}(P_{i,H_s}, \tilde{c}, t_s) \leq H_{i,s}^u$, $\forall \tilde{c} \in \mathscr{W}_i(t)$, where $H_{i,s}^u = -\varepsilon \ln \varepsilon - (1 - \varepsilon) \ln(1 - \varepsilon) > 0$ is a preset threshold of some small value.

For every vehicle $\mathscr{V}_i$, the motion control scheme with memory is given as follows:

$$\mathbf{u}_i^*(t) = \begin{cases} \bar{\mathbf{u}}_i(t) & \text{if Condition C6 does not hold} \\ \bar{\bar{\mathbf{u}}}_i(t) & \text{if Condition C6 holds} \end{cases} \tag{2.50}$$

where

$$\bar{\mathbf{u}}_i(t) = \bar{k}_i \sum_{\tilde{c} \in \mathscr{W}_i(t)} \left(\left[(2P_i(X(\tilde{c}) = 1;t) - 1)^2 - 1 \right]^2 \cdot \underbrace{\sum_{\tau=0}^{t} \left(\beta^i(\tau+1) - \beta^i(\tau) \right)}_{\text{Memory Term}} \right),$$

is the nominal control law, where both the current probability of object presence $P_i(X(\tilde{c}) = 1;t)$ and the sensing capability β^i up to the current time step are used, and the perturbation control law chooses the centroid $\tilde{\mathbf{q}}_i^*$ of cell $\tilde{c}_i^*$ from the set $\mathscr{Q}_i(t) = \{\tilde{c} \in \mathscr{D} : H_{i,s}(P_{i,H_s}, \tilde{c}, t_s) > H_{i,s}^u\}$, which is based on the uncertainty information at the current time step and only available to vehicle $\mathscr{V}_i$ itself.

Simulation-Based Performance Comparison

Now a set of numerical simulations are provided to illustrate and compare the performances of both motion control schemes. Assume a square domain $\mathscr{D}$ with size 50×50, and discretize it into 2500 unit cells. The parameter M_i of the vehicle sensor is set as 0.4, which gives the highest value for β^i as 0.9, i.e., there is 90% chance that the sensor is sensing correctly at the location of the vehicle. The sensing capability gradually decreases to $b_n = 0.5$. The desired uncertainty level is $H_{i,s}^u = 0.02$, corresponding to $\varepsilon = 0.0002$. There are 10 objects with a randomly selected deployment as indicated by the magenta dots in Figure 2.21(a). The position and radius for each of the 6 vehicle sensors is shown by the black dot and circle.

Figure 2.21(b) shows the probability of object presence according to vehicle $\mathscr{V}_1$ at time step $t = 1200$ under both control schemes. All the peaks represent the position of the objects detected with probability 1. The probability of object presence as estimated by other vehicles is similar to that shown in Figure 2.21(b). This indicates that all the unknown objects of interest have been found.

Figure 2.22(a) shows the trajectories of all the vehicles during the entire mission under the motion control scheme without memory. The green dots represent for vehicles' initial positions and red dots for final positions. Figure 2.22(b) shows the trajectories of all the vehicles under the motion control scheme with memory.

Figure 2.23(a) shows the the cost function $\mathscr{J}_i(t)$ for vehicles $\mathscr{V}_1$ to $\mathscr{V}_6$, respectively under the motion control scheme without memory. Figure 2.23(b) shows the the cost function $\mathscr{J}_i(t)$ under the motion control scheme with memory. Here the control law in Equation (2.50) is used with control gain $\bar{k}_i = 1, \bar{\bar{k}}_i = 0.025$. In both cases, all the cost functions converge to zero at time step $t = 1200$, which is consistent with the result shown in Figure 2.21(b) and equivalent to the detection of all the 10 unknown objects of interest.

Comparing the simulation results, there is more redundancy in vehicle trajectories under the memoryless motion control scheme. This is because the controller is only dependent on the current uncertainty map and does not take into account the history of the paths that the vehicles traveled before. However, the reduction of uncertainty is faster under the memoryless control scheme because it is a global

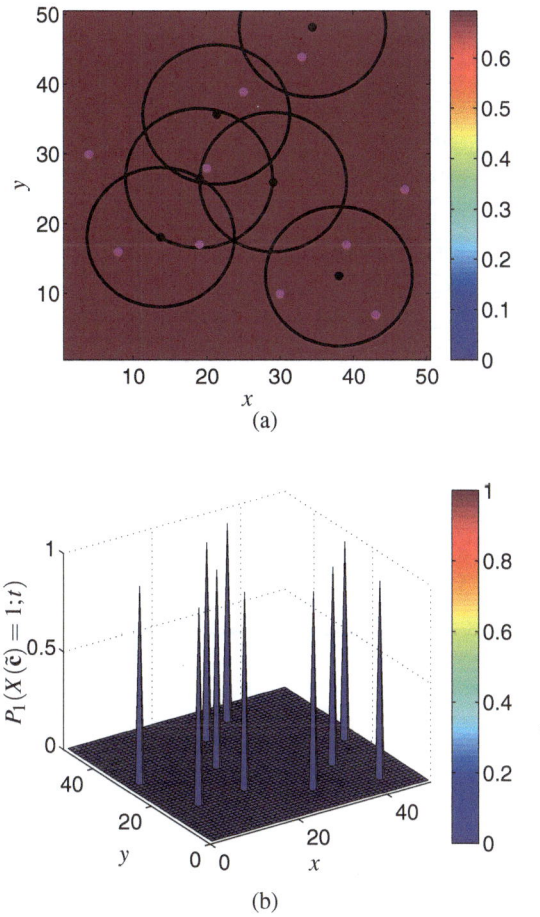

Fig. 2.21 Deployment of objects and vehicles, and probability of object presence.

controller that always seeks the cell with highest uncertainty within the entire search domain. On the other hand, under the motion control scheme with memory, the nominal controller is a local controller which drives the vehicle towards the cell with higher uncertainty within the sensory domain, and a perturbation controller is used whenever the vehicle is trapped in a local minimum. Under both motion control schemes, all the unknown objects of interest are found with desired uncertainty level. If fuel efficiency is a priority, one may want to avoid using a memoryless motion controller that spreads all over the domain. On the contrary, if time is a limited resource, one may prefer a memoryless motion controller in order to achieve the desired detection certainty quicker.

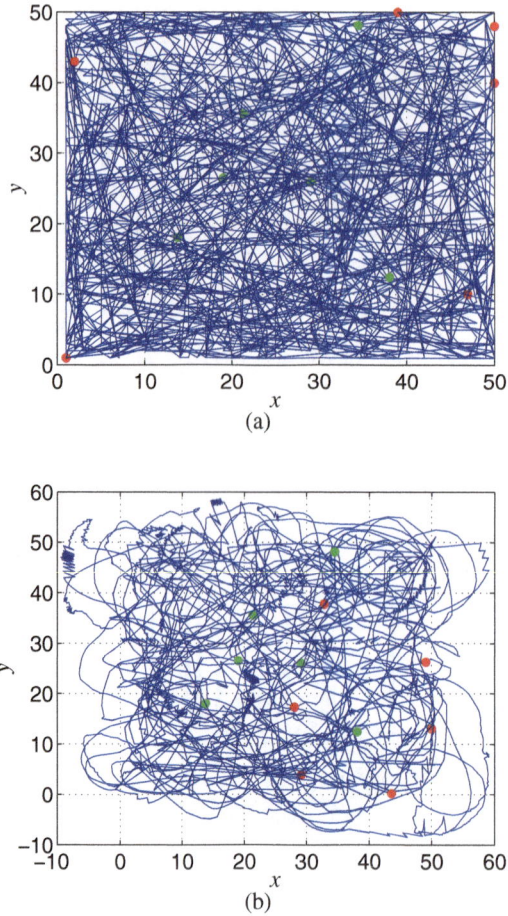

Fig. 2.22 Fleet motion under search control scheme without and with memory.

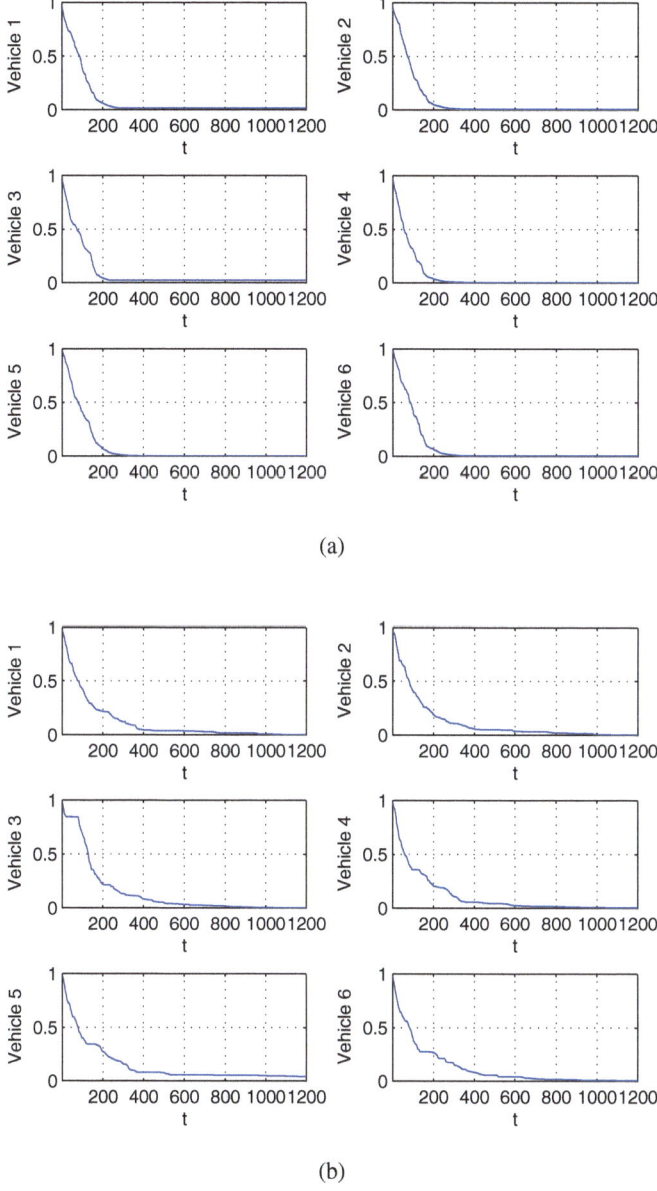

(a)

(b)

Fig. 2.23 Cost function under motion control scheme without and with memory.

Chapter 3
Awareness-Based Decision-Making Strategy

3.1 Problem Setup

In the search task, all objects of interest in a search domain are required to be found. In the classification task, each found object has to be classified for at least an amount of time equal to τ_c, which is the critical minimum information collection time that is needed to characterize the state of an object. Here the objects are assumed to be static.

Let $N_o \geq 0$ be the number of objects to be found and classified. Both N_o and the locations of the objects in $\mathscr{D}$ are unknown beforehand. At time t, let the set $\mathscr{A} = \mathscr{S}(t) \cup \mathscr{T}(t) = \{1,\ldots,N_a\}$, which is the set of indices of all the vehicles in the sensor fleet, and where the set $\mathscr{S}(t)$ contains indices of vehicles carrying out the search mission, and where the set $\mathscr{T}(t)$ contains indices of vehicles carrying out an object classification mission. Here assume that vehicles can either be searching or classifying at any instant time t, but not both simultaneously, and therefore the sets $\mathscr{S}(t)$ and $\mathscr{T}(t)$ are disjoint for all t. Initially, assume that all vehicles are in $\mathscr{S}(t)$. When a search vehicle detects an object and decides to classify its property, this search vehicle turns into a classification vehicle and, hence, there is one fewer vehicle in the set $\mathscr{S}(t)$ and one more vehicle in the set $\mathscr{T}(t)$.

Assuming some search versus classification decision-making strategy that guarantees coverage of the entire domain and that avoids the assignment of multiple vehicles to the classification of a single object, for the case when $N_o \leq N_a$, after a certain amount of time, each object will be guaranteed to be detected and its property satisfactorily classified by some vehicle. However, for the worst case scenario where $N_o > N_a$ in a large-scale domain and with a poor choice of decision-making strategy, one may end up with $\mathscr{S}(t) = \varnothing$ while there may still exist unfound objects. For example, a strategy where once an object is found it is classified for all time from that point forward would likely lead to some objects never being detected when there are more objects than vehicles. This section investigates strategies that guarantee

Y. Wang and I.I. Hussein: Search and Classification Using MAV, LNCIS 427, pp. 69–78.
springerlink.com   © Springer-Verlag London Limited 2012

that each object will be found and classified, especially under the worst case scenario, while simultaneously providing a lower bound for the amount of classification time.

It is assumed that each vehicle $\mathcal{V}_i \in \mathcal{A}$ satisfies the awareness dynamics (2.17). The state of awareness of the set of search agents $\mathscr{S}(t)$ in surveying $\bar{\mathbf{q}}$ then satisfies the differential equation (2.18).

3.2   Search and Classification Metrics

Similar to the probabilistic counterpart (2.47) proposed for MAV search mission in Section 2.4.5.3, here, the cost associated with a decision not to carry out further searching, $\mathscr{J}_1(t)$, is chosen to be proportional to the size of the un-searched domain. A uniform probability distribution is assumed for the locations of objects in $\mathscr{D}$, hence, $\mathscr{J}_1(t)$ is proportional to the probability of finding another object beyond time t. The cost associated with a decision not to classify found objects, $\mathscr{J}_2(t)$, is chosen to be proportional to the time spent not classifying a found object.

Define the search cost function to be

$$\mathscr{J}_1(t) = \frac{e_g(t)}{e_{g,\max}}, \tag{3.1}$$

where $e_g(t)$ is given by Equation (2.21). Under Assumption 2.3.1 and considering a uniform probability distribution for the locations of the objects in $\mathscr{D}$, the maximum value of $e_g(t)$ is given by

$$e_{g,\max} = e_g(0) = \frac{A_\mathscr{D}}{2}$$

because $x_{i0} = -1$. According to this definition, it follows that $0 \leq \mathscr{J}_1(t) \leq 1$. Initially, $\mathscr{J}_1(0) = 1$ describes the fact that it is known with probability 1 that there exists at least one object which has not been detected. This comes from the assumption that $N_o > 0$. If N_o happens to be zero, assuming that there exists at least one object of interest over the domain will guarantee verifying that there is none. Under Assumption 2.3.1, when $\mathscr{J}_1(t_s) = 0$ for some time $t_s > 0$, the entire domain has been satisfactorily covered and it is sure that there are no objects yet to be found. At this point, the search process is said to be completed.

For the classification metric $\mathscr{J}_2(t)$, let $\bar{N}_o(t) \leq N_o$ be the number of objects found by the sensor fleet up to time t. Define the classification cost function $\mathscr{J}_2(t)$ to be

$$\mathscr{J}_2(t) = \int_0^t \sum_{j=1}^{\bar{N}_o(\tau)} g_j(\tau)\mathrm{d}\tau, \tag{3.2}$$

where

$$g_j(t) = \begin{cases} 1 \text{ if } \mathbf{p}_j(t) \notin \mathscr{W}_i(t) \text{ for all } i \in \mathscr{A} \\ 0 \text{ if } \mathbf{p}_j(t) \in \mathscr{W}_i(t) \text{ for some } i \in \mathscr{A}. \end{cases}$$

If a search vehicle detects an object $\mathscr{O}_j$ a function $g_j(t)$ is assigned to the object (unless it has already been assigned one if detected in the past). A value of 0 is assigned to g_j as long as some agent classifies $\mathscr{O}_j$, and the classification cost associated with $\mathscr{O}_j$ is zero. In this case, $\mathscr{O}_j$ will be labeled as "assigned". Once the search vehicle decides not to classify $\mathscr{O}_j$, $\mathscr{O}_j$ is now labeled "unassigned", and $g_j(t)$ switches its value to 1, implying that a cost is now associated with not classifying the found object $\mathscr{O}_j$. According to Equation (3.2), this cost is equal to the amount of time during which a found object is not classified.

Remark 3.2.1. *A remark on the case with some information loss. If relaxing Assumption 2.3.1, the parameter ζ in the awareness model reflects loss of spatial information over time. It essentially sets a periodicity to how often the entire area must be re-surveyed. On the other hand, g_j reflects loss of information associated with a specific object over time. It is important to realize this distinction between the domain-awareness loss nature of ζ (and, hence, $\mathscr{J}_1$) and the specific-object awareness loss nature of g_j (and, hence, $\mathscr{J}_2$).* •

3.3 Search versus Classification Decision-Making

Under Assumption 2.3.1, a search and classification decision-making strategy will be developed to guarantee, in both its centralized and decentralized implementations, finding all objects in $\mathscr{D}$ and classifying each object for some time with a lower bound on the classification time.

3.3.1 Centralized Strategy

Since it is assumed that $N_o > N_a$, whenever a vehicle detects an object, it has to decide whether to classify it or to continue searching. If it does decide to classify, it has to decide on how much time it can afford to classify before it continues the search process.

Before deriving one possible way to determine the amount of classification time, first consider a search strategy. The goal in the search strategy is to attain an awareness level of $\|\mathbf{x}(\tilde{\mathbf{q}},t)\| \leq \xi$ for all $\tilde{\mathbf{q}} \in \mathscr{D}$ and all $t \geq t_s$ for some $t_s > 0$. For the search process, the control law (2.27) is used to drive the state of lack of awareness to a neighborhood of zero. It guarantees coverage of the entire domain $\mathscr{D}$ with $\mathscr{J}_1(t)$ converging to a small neighborhood of zero, which implies that all objects have been found and the search process is complete. The classification strategy discussed below will guarantee that all objects will be classified for a minimum of τ_c amount of time. The search control law (2.27) and the tracking strategy, together, will guarantee the detection of all objects of interest and their classification for at least τ_c amount of time.

If a search vehicle finds object(s) within its sensory range, then it will classify the object(s) for a T time period from the time of detection, where

$$T = \frac{\tau_c}{\mathscr{J}_1(t_d)}, \tag{3.3}$$

t_d being the time of object detection, and where $\tau_c > 0$ is the desired critical minimum amount of classification time. This is the amount of time that is needed to characterize the property of an object. The larger the value of $\mathscr{J}_1(t_d)$ is (i.e., the less aware the vehicle is of the domain), the less time the vehicle will spend classifying the object. As the degree of awareness increases at detection time, the more time the vehicle spends classifying the object. Note that $\mathscr{J}_1(t_d)$ can not be zero unless the mission is completed, at which point there is no need to compute T.

Hence, once a vehicle detects an object and decides to classify this particular object, it becomes a classification vehicle and will not carry out any searching for a period of T seconds. Note that while the vehicle is classifying, other vehicles may be searching. In the centralized implementation, the amount of centralized system awareness $\mathbf{x}(\tilde{\mathbf{q}},t)$ is available to all vehicles. So is the value of $\mathscr{J}_1(t_d)$. It is assumed that each object will only be classified once by only one vehicle during the mission. After a time period of T, the classification vehicle will switch back to become a search vehicle and leave its classification position to find new objects. At this point in time, the object will be labeled "assigned" and will not be classified by any other vehicle if found.

Theorem 3.3.1. *Under Assumption 2.3.1, the centralized search and classification decision-making strategy given by Equations (2.27) and (3.3) will guarantee that $\mathscr{J}_1$ converges asymptotically to zero, which is equivalent to guaranteeing that all objects be found. The minimum amount of time spent classifying any object is given by τ_c.*

Proof. The proof for guaranteed detection of all object follows directly from Theorem 2.3.4.

The minimum classification time comes from the fact that once an object is found, it will be classified for at least $\tau_c/\mathscr{J}_1(t_d)$. $\mathscr{J}_1(t_d)$ assumes a maximum value of 1 if $t_d = 0$. In the extreme scenario where an object is found at $t = 0$, the value of T is exactly τ_c. If an object is found at a time other than $t = 0$, $\mathscr{J}_1(t_d)$ has to be less than 1 and, hence, T is greater than τ_c. $\square$

Remark 3.3.1. *For the case when N_o is known beforehand and $N_o \leq N_a$, under the centralized search, and assuming that if some vehicle finds an object it will classify this object for all future time, each object will be guaranteed to be detected and its property permanently classified by some vehicle. Proof of complete coverage of*

Table 3.1 Classification time T for each object.

	Object 1	Object 2	Object 3	Object 4	Object 5	Object 6
$T,(s)$	8.0583	50.2437	7.5215	5.2552	10.3786	6.6144

the domain, and, hence, detection of each object, follows directly from the proof of Theorem 3.3.1. Since $N_o \leq N_a$ and each object can only be classified by one vehicle, assigning a unique vehicle to a single object whenever an object is detected is feasible (i.e., there are enough resources to do so) and every object will be satisfactorily classified.

A simulation result is provided in Figures 3.1 and 3.2, where $N_o = 6$ and $N_a = 4$ for some choice of controller gains and coverage sensor parameters. The domain $\mathcal{D}$ is square in shape and discretized into $N_{tot} = n \times n = 32 \times 32$ cells, where $\tilde{\mathbf{q}} \in \mathbb{R}^2$ represents the centroid of each cell. Hence, $\mathbf{x}(\tilde{\mathbf{q}},t)$ can be written as a vector of dimension $2n$. Figures 3.4(a) and 3.4(b) show the evolution of $\mathcal{J}_1(t)$ and $\mathcal{J}_2(t)$ under this centralized control strategy. Figures 3.1(c) and 3.1(d) show the control force and fleet motion under the centralized implementation. Figure 3.2 shows the state of awareness distribution at three different time instances. The circular dots indicate the positions of the vehicles, and the square dots indicate the objects. The magenta circles are the vehicles' sensor ranges. Table 3.1 shows the classification time of each object, which is guaranteed to be at least $\tau_c = 5$ seconds.

3.3.2 Decentralized Strategy

Now assume that the sensor fleet is completely decentralized. That is, each vehicle is aware of coverage achieved by itself alone. Each object it finds will be assumed to be found for the first time. This represents a scenario where communications between vehicles is not possible (for example, due to security reasons, the sensor vehicles have to remain "silent" otherwise they themselves may be detected by adversary vehicles).

In the decentralized formulation, the search control strategy (2.26) is employed. For the classification strategy, when a search vehicle $\mathcal{V}_i$ finds object(s) within its sensory range, it classifies the objects for a time period of T, defined by

$$T = \frac{\tau_c}{\mathcal{J}_{1i}(t_d)} \tag{3.4}$$

where

$$\mathcal{J}_{1i}(t) = \frac{e_{gi}(t)}{e_{gi,max}}, \tag{3.5}$$

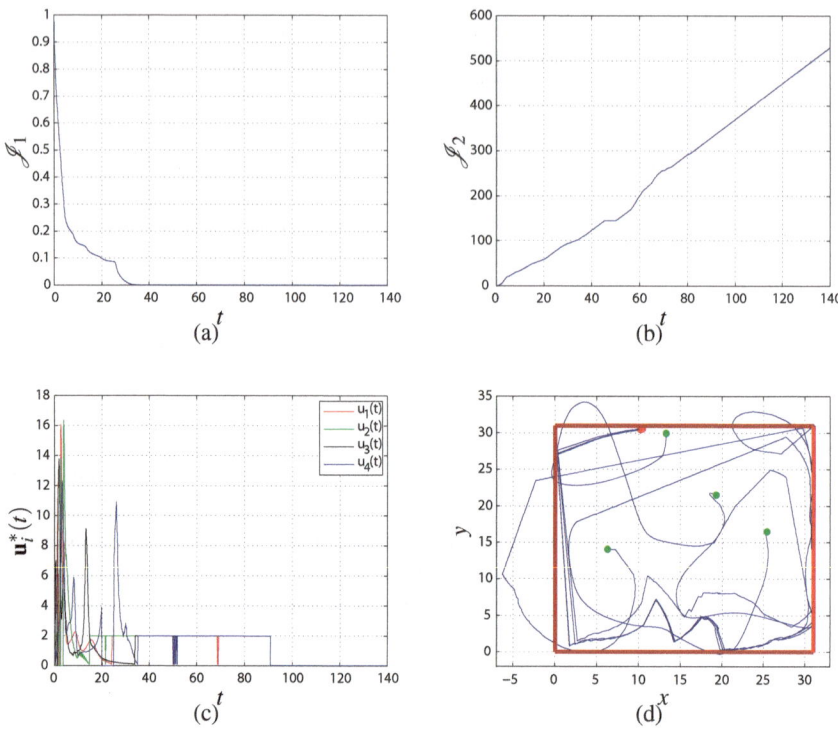

Fig. 3.1 Centralized implementation (awareness-based decision-making).

and where $e_{gi}(t)$ (Equation (2.20)) is the global error over the entire mission domain achieved by the vehicle $\mathcal{V}_i$ only, with $e_{gi,\max} = e_{gi}(0)$ being half of the area of $\mathcal{D}$ if the initial state $\mathbf{x}_i(\tilde{\mathbf{q}}, t = 0) = -1$ is as assumed from the outset. Moreover, define the cost of not classifying an object found by vehicle $\mathcal{V}_i$ by

$$\mathscr{J}_{2i}(t) = \int_0^t \sum_{j=1}^{\bar{N}_0^i(\tau)} g_j(\mathbf{p}_j(\tau)) d\tau, \qquad (3.6)$$

where $\bar{N}_0^i(t)$ is the number of objects found by vehicle $\mathcal{V}_i$ up to time t. Assume that each object will only be classified once by each vehicle during the mission.

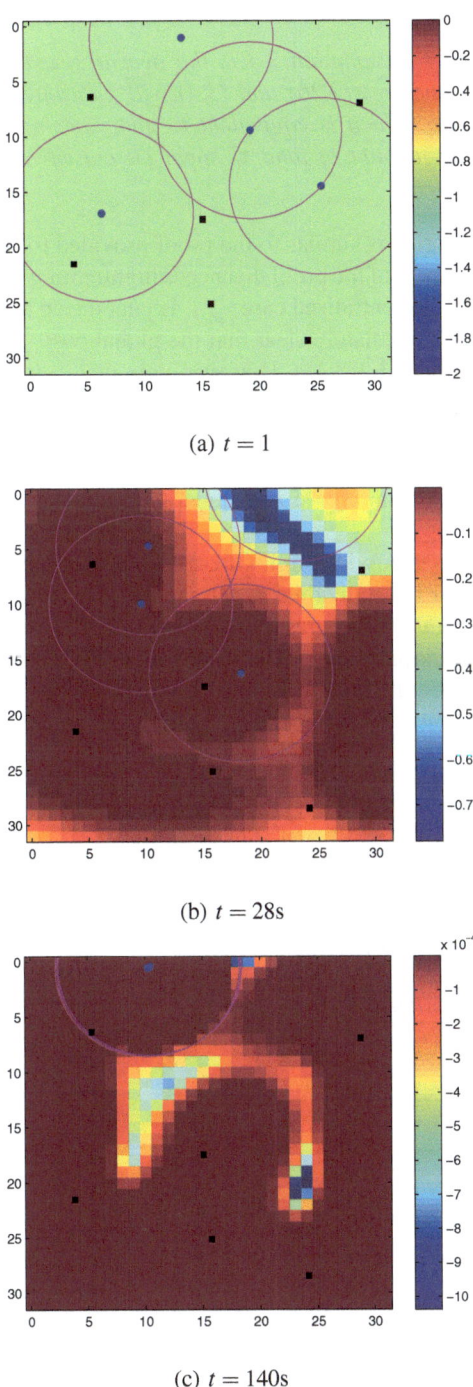

(a) $t = 1$

(b) $t = 28$s

(c) $t = 140$s

Fig. 3.2 State of awareness at different time instances (Centralized).

Similar to Theorem 3.3.1, the following results hold:

Theorem 3.3.2. *Under Assumption 2.3.1, the decentralized search and tracking strategy given by Equations (2.26) and (3.4) will guarantee that $\mathcal{J}_1$ converges asymptotically to zero, which is equivalent to guaranteeing that all objects has found. The minimum amount of time τ_c spent on classifying any object is also achieved by each vehicle.*

The proof of this theorem is similar to the proof provided for the centralized case. The only important aspect of the proof that needs highlighting is that, along the same lines as the proof for the centralized case, $\mathcal{J}_{1i}$ is guaranteed to converge to zero for all $\mathcal{V}_i \in \mathcal{A}$. It is not immediately clear that the global cost $\mathcal{J}_1$ will also converge to zero as the Theorem 3.3.2 states. However, note that $e_{gi}(t) \geq e_g(t)$ because the more vehicles and sensors available to us, at least the same or higher overall global coverage is achieved by the system. Since $\mathcal{J}_1$ and $\mathcal{J}_{1i}$ (for all $\mathcal{V}_i \in \mathcal{A}$) are both initialized to be 1, then $\mathcal{J}_{1i}(t) \geq \mathcal{J}_1(t)$, for all time t, because $e_{gi}(t) \geq e_g(t)$. If $\mathcal{J}_{1i}(t)$ is guaranteed to converge to zero under the control law (2.26), then so does $\mathcal{J}_1(t)$.

A simulation result is provided in Figures 3.3 and 3.4. Figures 3.3(a) and 3.3(b) show the evolution of the individual $J_{1i}(t)$ and $J_{2i}(t)$, $i = 1,2,3,4$ under the decentralized control strategy. Figures 3.3(c) and 3.3(d) show the control force and fleet motion under the decentralized implementation. Figure 3.4 shows the state of awareness distribution at three different time instances. Table 3.2 shows the classification time of each object by each vehicle, which is guaranteed to be at least $\tau_c = 5$ seconds.

Table 3.2 Classification time of each object by each vehicle.

$T, (s)$	Object 1	Object 2	Object 3	Object 4	Object 5	Object 6
Agent 1	7.0090	5.8723	5.1221	5.3971	6.1709	5.6022
Agent 2	5.5974	7.0428	5.1835	5.0000	5.6022	6.1474
Agent 3	8.7574	7.9469	5.1609	5.6027	5.3631	5.1281
Agent 4	5.7563	5.1835	7.0981	6.3030	5.9109	6.5911

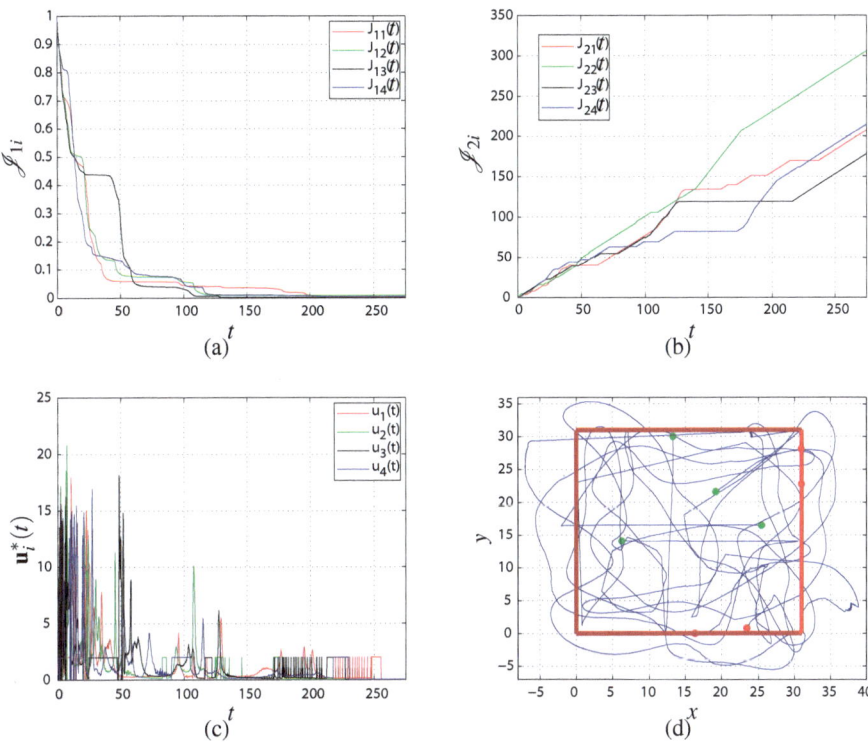

Fig. 3.3 Decentralized Implementation (awareness-based decision-making).

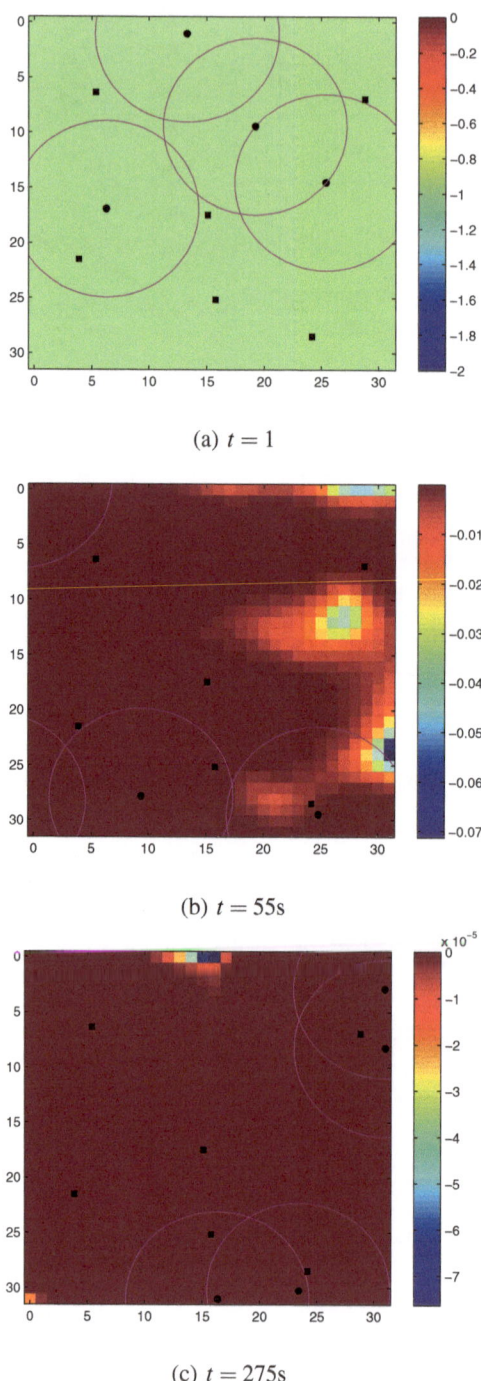

(a) $t = 1$

(b) $t = 55s$

(c) $t = 275s$

Fig. 3.4 State of awareness at different time instances (Decentralized).

Chapter 4
Bayesian-Based Decision-Making Strategy

4.1 Problem Setup

To illustrate the ideas while avoiding additional computation complexities, in this and the subsequent chapters, it is assumed that there is a single autonomous vehicle performing the search and classification tasks under the probabilistic frameworks. This reflects the case of extremely limited sensing resources, i.e., a single autonomous vehicle as opposed to cooperative MAVs. The extension to MAV decision-making can follow the formulation presented in Section 2.4.5 via sensor fusion. Section 5.4 in Chapter 5 discusses the extended application of risk-based sequential decision-making to the Space Situational Awareness (SSA) problem using a Space-Based Space Surveillance (SBSS) system, which consists of both ground-based sensors and orbiting satellites.

For both the search and classification processes, the Bernoulli-type limited-range sensor model (2.31,2.32) in Section 2.4.1 is used, however, with different observation contents: $X(\tilde{\mathbf{c}}) = 0$ for object "present" and $X(\tilde{\mathbf{c}}) = 1$ for object "absent" in search, and $X_c(\mathbf{p}_k) = 0$ for object $\mathscr{O}_k$ having property 'F' and $X_c(\mathbf{p}_k) = 1$ for object $\mathscr{O}_k$ having property 'G' in classification. Here, an object can be assigned as many property types as needed, but without loss of generality, it is assumed that an object can have one of two properties, either Property 'F' or Property 'G'. Let $Y_c(\mathbf{p}_k)$ be the corresponding classification observation random variable, where $Y(\mathbf{p}_k) = 0$ corresponds to the observation indicating that there is an object $\mathscr{O}_k$ with property 'F' present at position $\mathbf{p}_k$ and $Y(\mathbf{p}_k) = 1$ corresponds to property 'G', respectively. The actual observation is taken according to the probability parameter β_c of the Bernoulli distribution. The general conditional probability matrix B_c for the classification process is then given as follows

$$B_c =$$
$$\begin{bmatrix} \text{Prob}(Y_c(\mathbf{p}_k) = 0 | X_c(\mathbf{p}_k) = 0) = \beta_c & \text{Prob}(Y_c(\mathbf{p}_k) = 0 | X_c(\mathbf{p}_k) = 1) = 1 - \beta_c \\ \text{Prob}(Y_c(\mathbf{p}_k) = 1 | X_c(\mathbf{p}_k) = 0) = 1 - \beta_c & \text{Prob}(Y_c(\mathbf{p}_k) = 1 | X_c(\mathbf{p}_k) = 1) = \beta_c \end{bmatrix}$$

Y. Wang and I.I. Hussein: Search and Classification Using MAV, LNCIS 427, pp. 79–88.
springerlink.com

Similar as in Section 2.4.1, two types of sensor models can be assumed for classification. For the unit-range sensor model, β_c is set as a constant value. For the limited circular range sensor model, the following example is in a same fashion as Equation (2.32),

$$\beta_c(s) = \begin{cases} \frac{M_c}{r_c^4}\left(s^2 - r_c^2\right)^2 + b_n & \text{if } s \le r_c \\ b_n & \text{if } s > r_c \end{cases},\qquad(4.1)$$

where $M_c + b_n$ is the maximum sensing capability, $s = \|\mathbf{q}(t) - \mathbf{p}_k\|$, $k = 1, 2, \cdots, N_o$, and r_c is limited classification sensory range. When an object of interest is within the sensor's effective classification radius $\tilde{r}_c < r_c$, this object is said to be found, and the vehicle has to decide whether to classify it or continue searching. Bayes' rule is employed to update the probability of object presence at cell $\tilde{\mathbf{c}}$ for the search process. Similar as Equations (2.34) and (2.33), we use Bayes rule to update the probability of a found object $\mathcal{O}_k$ having property 'G' for the classification process, i.e., $P_c(X_c(\mathbf{p}_k) = 1)$. Based on the updated probability of object existence, define an information entropy function $H_s(P_{H_s}, \tilde{\mathbf{c}}, t)$ (2.35) as a measure of uncertainty for the search process. For the classification process, define a similar information entropy function $H_c(P_{H_c}, \mathbf{p}_k, t)$ as Equation (2.35) for every found object $\mathcal{O}_k$ to evaluate its classification uncertainty:

$$\begin{aligned}&H_c(P_{H_c}, \mathbf{p}_k, t)\\ &= -P_c(X_c(\mathbf{p}_k) = 0)\ln P_c(X_c(\mathbf{p}_k) = 0) - P_c(X_c(\mathbf{p}_k) = 1)\ln P_c(X_c(\mathbf{p}_k) = 1),\end{aligned}$$

where the probability distribution P_{H_c} for the classification process is given by $P_{H_c} = \{P_c(X_c(\mathbf{p}_k) = 0), P_c(X_c(\mathbf{p}_k) = 1)\}$. There are as many scalar H_c's as there are found objects $\mathcal{O}_k$ up to time t. The initial value for H_c for every found object $\mathcal{O}_k$ can also be set as $H_c = H_{c,\max} = 0.6931$.

4.2 Task Metrics

This section develops metrics to be used for the search versus classification decision-making process. For the search process, a same metric as Equation (2.47) is presented when applied to a single vehicle sensor. In the event of object detection and a decision not to proceed with the search process, but, instead, to stop and classify the found object, the associated cost is defined as

$$\mathcal{J}(t) = \frac{\sum_{\tilde{\mathbf{c}} \in \mathcal{D}} H_s(P_{H_s}, \tilde{\mathbf{c}}, t)}{H_{s,\max} A_{\mathcal{D}}}.\qquad(4.2)$$

For the classification process, let $\bar{N}_o(t)$ be the number of objects found by the autonomous sensor vehicle up to time t. For each found object $\mathcal{O}_k \in \{1, 2, \cdots, \bar{N}_o(t)\}$, define the classification metric $H_d(\mathbf{p}_k, t)$ to be

$$H_d(\mathbf{p}_k, t) = H_c^u \mathcal{J}(t), \qquad (4.3)$$

where the weighting parameter $H_c^u \in (0, 1)$ is a preset upper bound on the desired uncertainty level for classification. This metric couples the search and classification processes and allows decision-making based on the real-time progress of the mission. H_d depends on how uncertain the vehicle is of the presence of more unfound objects in $\mathcal{D}$ through $\mathcal{J}$. If the vehicle finds an object $\mathcal{O}_k$ (i.e., within the effective classification radius $\bar{r}_c$) and decides to classify it, the vehicle will continually classify it and compare the classification uncertainty $H_c(P_{H_c}, \mathbf{p}_k, t)$ to the desired classification uncertainty $H_d(\mathbf{p}_k, t)$. Only when the classification condition

$$H_c(P_{H_c}, \mathbf{p}_k, t) < H_d(\mathbf{p}_k, t_d) \qquad (4.4)$$

is satisfied, the vehicle stops classifying the found object and switch to search again. Here, t_d is the time of object detection.

The highest classification uncertainty bound H_c^u is motivated by the following. Say that the vehicle detects an object at the beginning of the mission with $t_d = 0$ and decides to classify it. Initially, $\mathcal{J}(0) = 1$ and the vehicle will attempt to classify it until $H_c < H_c^u$. This is the minimum desired classification accuracy level for any found object. Any further classification accuracy will come at the cost of not performing the search task and decrease the potential of finding more critical objects in the domain. If an object is found at a time other than $t_d = 0$, $\mathcal{J}(t_d)$ has to be less than 1 and, hence, $H_d(\mathbf{p}_k, t_d)$ is smaller than H_c^u. On the other end of the spectrum, if $\mathcal{J}(t_d) = 0$, the vehicle can spend as much time classifying the object since it does not come at any search cost. This is because the vehicle has achieved 100% certainty that it has found all critical and noncritical objects in the domain.

4.3 Search vs. Classification Decision-Making

Now consider a probabilistic Bayesian-based search versus classification decision-making strategy that guarantees finding all the unknown objects in $\mathcal{D}$ (i.e., achieve $\mathcal{J} \to 0$) and classifying each object with an upper bound H_c^u of the classification uncertainty.

For the search strategy, the control law (2.36) is used and the following lemma holds.

Lemma 4.3.1. *Assume $\mathcal{U}$ is such that $\mathcal{D} = \mathcal{Q}_{\mathcal{D}}(t)$, the search strategy (2.36) guarantees an uncertainty level $H_s(P_{H_S}, \tilde{\mathbf{c}}, t_s) \leq H_s^u$ for all $\tilde{\mathbf{c}} \in \mathcal{D}$. Therefore, the search cost $\mathscr{J}(t) \leq \varepsilon_s = \frac{H_s^u}{H_{s,\max}}$ for all $t \geq t_s$ for some $t_s > 0$. This is equivalent to the detection of all unknown objects in $\mathcal{D}$ with a desired certainty level.*

Proof. If Condition **C5** does not hold, the nominal control law $\bar{\mathbf{u}}(t)$ is utilized to drive the vehicle to some cell $\tilde{\mathbf{c}}_\star$ that has the highest search uncertainty in $\mathcal{Q}_{\mathcal{W}}(t)$.

When the uncertainty H_s of all the cells $\tilde{\mathbf{c}} \in \mathcal{Q}_{\mathcal{W}}(t)$ converges to H_s^u, Condition **C5** holds, and the vehicle gets trapped in regions of $H_s \leq H_s^u$ by applying only the nominal control law $\bar{\mathbf{u}}$ while the entire domain $\mathcal{D}$ has not been fully searched yet.

At this moment, the perturbation control law $\bar{\bar{\mathbf{u}}}$ is used to drive the vehicle out of the regions with low uncertainty H_s^u to some $\tilde{\mathbf{c}}^* \in \mathcal{Q}_{\mathcal{D}}(t)$ such that $H_s(P_{H_S}, \tilde{\mathbf{c}}^*, t) > H_s^u$ if such a point exists. Under the perturbation control law, $\|\mathbf{q} - \tilde{\mathbf{q}}^*\|$ will eventually be smaller than r and, hence, Condition **C5** will not hold. At this point in time, the control is switched back to the nominal control law. Note that $\bar{\bar{\mathbf{u}}}$ is always in $\mathcal{U}$ by definition of $\mathcal{Q}_{\mathcal{D}}(t)$.

Given that $\mathcal{Q}_{\mathcal{D}}(t) \subseteq \mathcal{D}(t)$ according to definition, if $\mathcal{U}$ is such that any $\tilde{\mathbf{q}} \in \mathcal{D}(t)$ is also in $\mathcal{Q}_{\mathcal{D}}(t)$, viz., $\mathcal{D} = \mathcal{Q}_{\mathcal{D}}(t)$, then every cell in $\mathcal{D}$ is reachable from where the sensor is. The switching between $\bar{\mathbf{u}}$ and $\bar{\bar{\mathbf{u}}}$ is repeated until whenever Condition **C5** holds there does not exist $\tilde{\mathbf{c}}^*$. The non-existence of such a $\tilde{\mathbf{c}}^*$ at some time $t_s > 0$ guarantees that $\mathscr{J}(t_s)$ is sufficiently close to zero. Because $H_s(P_{H_S}, \tilde{\mathbf{c}}, t_s)$ is smaller than H_s^u everywhere within $\mathcal{D}$, it follows that $\mathscr{J}(t_s) \leq \frac{H_s^u}{H_{s,\max}} = \varepsilon_s$ according to the search cost function (4.2). The search mission is then said to be completed. $\square$

Next, consider the following classification strategy: A sensor vehicle will stop searching and begins to classify an object whenever the object is within its effective classification range $\tilde{r}_c$. If the classification condition (4.4) is satisfied, the vehicle will switch back to become a search vehicle and leave its classification position to find new objects. The vehicle can resume classifying an object that has been detected and classified in the past if it finds it again during the search process.

Lemma 4.3.2. *The classification strategy guarantees that each found object in $\mathcal{D}$ will be classified with an upper bound uncertainty H_c^u.*

Proof. Once the vehicle finds object $\mathscr{O}_k$ within its effective classification range $\tilde{r}_c$ and decides to classify it, it switches to a classification task and will not carry out any search until $H_c(P_{H_C}, \mathbf{p}_k, t) < H_d(\mathbf{p}_k, t_d)$. After achieving at least the desired upper bound of classification uncertainty H_c^u, the vehicle switches back to search again. When the vehicle left the object, the classification uncertainty for this object will remain constant until the vehicle comes back to classify it when possible. At that time, the value of H_d will be smaller because more regions have been searched since the last time the vehicle has found the object. This process will be repeated until each object in $\mathcal{D}$ has a classification uncertainty of at most H_c^u, or equivalently, the classification task is completed. $\square$

Theorem 4.3.1. *According to Lemma 4.3.1 and 4.3.2, the search and classification decision-making strategy guarantees that $\mathcal{J}$ converges asymptotically to zero, which is equivalent to guaranteeing that all the unknown objects within the domain will be found. The maximum acceptable classification uncertainty H_c^u is achieved by every found object.*

Remark 4.3.1. *The priority of each task during the mission is based on the real-time progress, that is, the corresponding task metrics at each time instant. In the current setting, whenever the object is within a sensor vehicle's effective classification range $\tilde{r}_c$, the vehicle will begin to classify the object. At that moment, the classification task possesses higher priority. The vehicle will switch back to search again when the classification uncertainty H_c is less than the desired classification uncertainty level H_d, which is time-varying and depends on the search uncertainty level H_s at the detection time t_d according to the classification metric (4.3) and the search cost function (4.2). At this point in time, the search task is given a higher priority. Because H_s is decreasing with time, H_d also decreases. Therefore, a vehicle will be able to spend more time classifying a found object when more unknown objects have been found than at the outset of the mission. This can be interpreted as that more priority will be assigned to the classification task as time increases.* •

4.4 Simulation

This section provides A) a detailed numerical simulation that illustrates the performance of the decision-making strategy, and B) a Monte-Carlo simulation study to demonstrate the properties of the proposed algorithms. All the simulations are implemented by means of a 2.27-GHz, i3-350m processor with 4GB RAM, and Matlab®-compiled codes.

4.4.1 Simulation Example

Assume a square domain $\mathscr{D}$ with size 32×32 units length, thus the domain is discretized into 1024 cells. There are $N_0 = 5$ objects. Let objects 1, 3 and 5 have property 'F', and objects 2 and 4 have Property 'G', with a randomly selected initial deployment as shown by the green and magenta crosses, respectively, in Figure 4.1. Figure 4.1 shows the evolution of search uncertainty H_s (dark red for highest uncertainty and dark blue for lowest uncertainty) and the vehicle motion at $t = 1,250,475$ and 700. The maximum radius r of the search sensor is chosen to be 8 and the classification radius r_c is also chosen as 8, as shown by the magenta circle in Figure 4.1. The effective classification radius $\tilde{r}_c$ is set as 6 as shown by the green circle in the figure. The black dot represents the position of the vehicle. The parameter $M = M_c$ of the sensor is set as 0.4, which gives the highest value for β as 0.9, i.e., there is 90% chance that the sensor is sensing correctly at the location of the vehicle.

The sensing capability gradually reduces to 0.5 according to the models discussed above (Equations (2.32) and (4.1)). The initial position of the vehicle is also selected randomly (see Figure 4.1(a)). Let the desired upper bound for classification uncertainty H_c^u be 0.01. Here the control law in Equation (2.36) is used with control gain $\bar{\bar{k}} = 0.2$. The set $\mathscr{U}$ is chosen to be $\mathscr{D}$, so that $\mathscr{Q}_{\mathscr{W}}(t)$ is given by the intersection of $\mathscr{U}$ and $\mathscr{W}$, i.e., $\mathscr{D} \cap \mathscr{W}$ and $\mathscr{Q}_{\mathscr{D}}(t) = \mathscr{D}$ which guarantees the full coverage of the entire domain. From Figure 4.1(d), it can be concluded that the desired zero search uncertainty has been achieved everywhere within $\mathscr{D}$. The actual maximum achieved search uncertainty turns out to be 2.6×10^{-3} according to the simulation results.

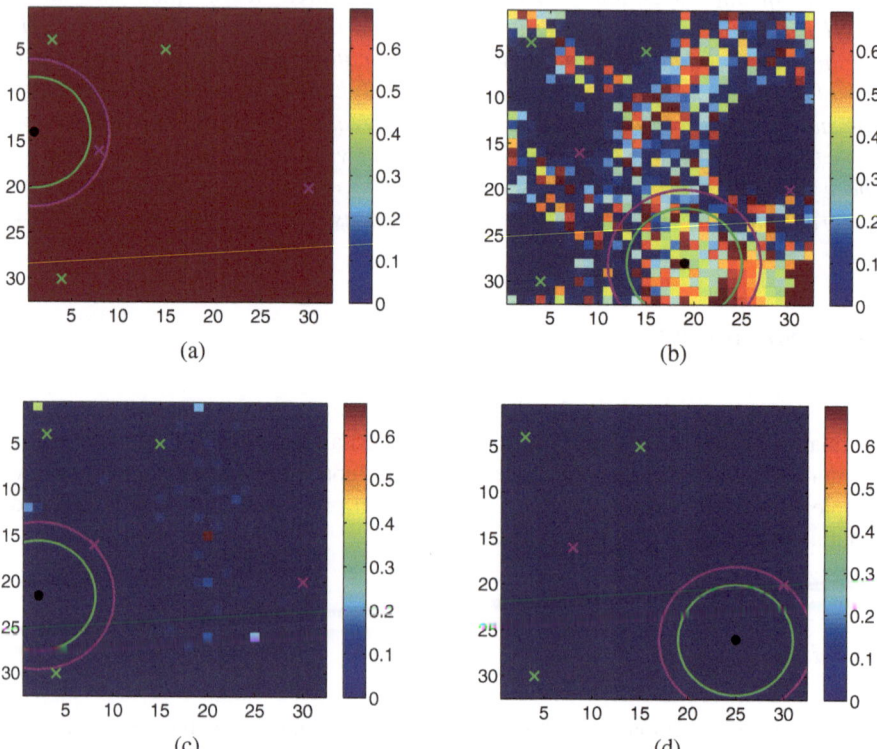

Fig. 4.1 Search uncertainty map (Bayesian-based decision-making).

Figure 4.2(a) shows the evolution of the search cost $\mathscr{J}(t)$ under the control strategy (2.36) and can be seen to converge to zero. All the objects have been found with the probabilities of object presence as 1 and zero search uncertainty. Those cells that do not contain an object end up with zero search probability and uncertainty. Figure 4.2(b) shows the posterior probabilities for every $\tilde{\mathbf{c}}$ within $\mathscr{D}$ at $t = 700$, where all the unknown objects are detected and all the empty cells are also identified.

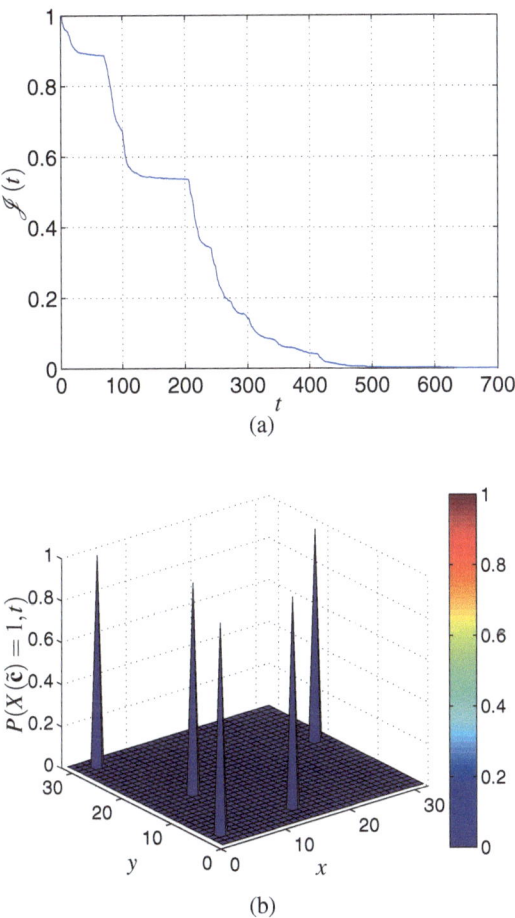

Fig. 4.2 Search cost function $\mathscr{J}(t)$ and posterior probabilities for search at $t = 700$.

For all the 5 found objects, objects $2, 4$ have been classified with probability of having Property 'G' as 1 and zero classification uncertainty. Objects $1, 3, 5$ have been classified with probability of having property 'G' as 0 and zero classification uncertainty. Figure 4.3 shows that, for example, object 2 has property 'G' and object 3 has property 'F' with zero classification uncertainty. The classification results of other objects can be shown like Figure 4.3 without difficulty.

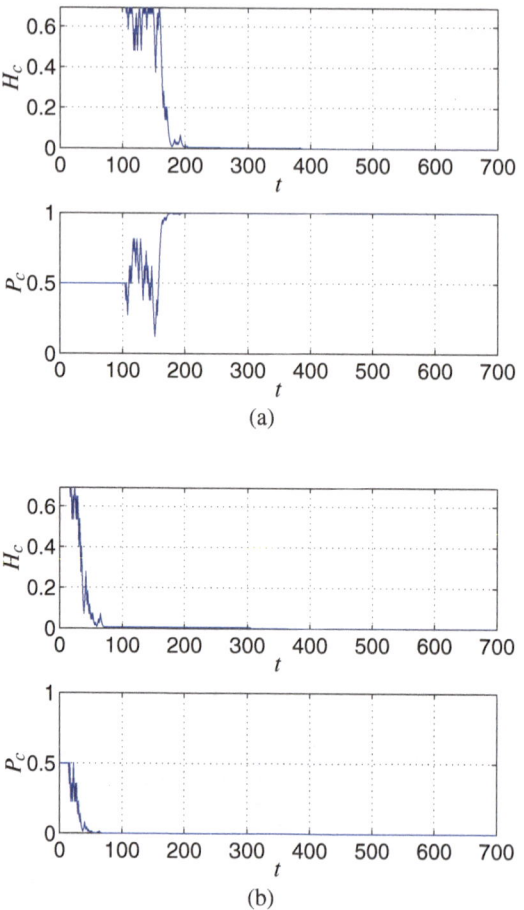

Fig. 4.3 Classification results for object 2 and 3 (Bayesian-based decision-making).

4.4.2 Monte-Carlo Simulation

In this section, a Monte-Carlo simulation-based study is provided to investigate the performance of the proposed strategy. Four metrics are used to evaluate the algorithms, i.e., the average CPU time for mission completion, the average simulation steps t for mission completion, the achieved mean search uncertainty over the domain, and the achieved mean classification uncertainty of all found objects and their corresponding standard deviations. The mission is said to be complete when a desired search and classification uncertainty of at most 0.01 has been achieved.

The algorithms is tested by varying the mission domain size, search and classification sensory ranges $(r, r_c, \tilde{r}_c)$, peak sensory capability M, and the total number of objects N_0. 100 runs are implemented for each case with a fixed combination of the above parameters. The statistical results are listed in Tables 4.1-4.4.

Table 4.1 shows the average CPU time for mission completion, the average simulation steps t for mission completion, the achieved mean search uncertainty $E[H_s]$ over the entire domain, and the achieved mean classification uncertainty $E[H_c]$ for all found objects with their corresponding standard deviations (in parentheses) of 100 runs under domain sizes 16×16, 24×24, 32×32, 40×40, respectively, using a fixed set of object positions under each case and same parameters as in Section 4.4.1. As expected, the time for mission completion grows with the domain size. An interesting observation is that as the domain size increases, the final achieved average search and classification uncertainty levels decreases. This is because in larger domains, more regions will have to be revisited in order to cover the entire domain. Moreover, note that the deviation of classification uncertainty is larger than the search uncertainty because every object is detected at a different time step and the corresponding H_d is time-varying.

Table 4.1 Varying mission domain size.

Size	CPU	t	$E[H_s]$	$E[H_c]$
16	1.33	311.72	2.73E-05	1.53E-04
	(0.66)	(120.67)	(9.16E-06)	(3.38E-04)
24	4.96	489.58	1.34E-05	1.17E-04
	(1.63)	(118.65)	(4.09E-06)	(4.52E-04)
32	13.25	695.53	8.13E-06	3.93E-05
	(3.29)	(113.22)	(1.94E-06)	(1.09E-04)
40	34.81	1043.4	5.76E-06	1.44E-05
	(6.74)	(121.77)	(2.08E-06)	(5.02E-05)

Table 4.2 shows the four metrics of 100 runs under sensory range $(r = r_c = 6, \tilde{r}_c = 4)$, $(r = 8, \tilde{r}_c = 6)$, $(r = 10, \tilde{r}_c = 8)$, $(r = 10, \tilde{r}_c = 7)$, respectively, using the same parameters as in Section 4.4.1. Since smaller sensory range is equivalent to larger domain size, it is expected that this case leads to more mission completion time. With the same search range, smaller classification range causes reduction in the mission completion time because the probability of object detection decreases with smaller sensory range. However, the final achieved uncertainty is higher under less mission completion time.

Table 4.3 shows the four metrics of 100 runs under peak sensory capability $M = M_c = 0.2$, $M = 0.3$, $M = 0.4$, $M = 0.5$, respectively, using the same parameters as in Section 4.4.1. The larger M is, the shorter the mission completion time and the lower the search uncertainty becomes. Note that when $M = 0.5$, the vehicle has perfect sensing, i.e., 100% detection probability, at its location. This leads to a big reduction in mission completion time and final achieved uncertainty.

Table 4.2 Varying sensory range.

Range	CPU	t	$E[H_S]$	$E[H_c]$
$r=6$	20.66	916.35	8.68E-06	4.00E-05
$\tilde{r}_c=4$	(3.07)	(64.33)	(1.28E-06)	(1.77E-04)
$r=8$	13.25	695.53	8.13E-06	3.93E-05
$\tilde{r}_c=6$	(3.29)	(113.22)	(1.94E-06)	(1.09E-04)
$r=10$	12.96	680.68	7.86E-06	3.01E-05
$\tilde{r}_c=8$	(4.26)	(166.89)	(2.00E-06)	(6.60E-05)
$r=10$	8.13	452.15	8.01E-06	4.47E-05
$\tilde{r}_c=7$	(1.44)	(57.39)	(1.67E-06)	(1.62E-04)

Table 4.3 Varying peak sensing capability.

M	CPU	t	$E[H_S]$	$E[H_c]$
0.2	96.31	2837.31	9.30E-06	2.82E-05
	(21.34)	(320.35)	(6.88E-07)	(1.99E-04)
0.3	30.37	1275.09	8.69E-06	1.74E-05
	(7.29)	(169.55)	(1.17E-06)	(4.62E-05)
0.4	13.25	695.53	8.13E-06	3.93E-05
	(3.29)	(113.22)	(1.94E-06)	(1.09E-04)
0.5	7.90	421.60	6.20E-06	2.58E-05
	(1.97)	(80.12)	(3.12E-06)	(3.69E-05)

Table 4.4 shows the four metrics of 100 runs under 3, 5, 10, 20 objects, respectively, using a fixed set of object positions under each case and the same parameters as in Section 4.4.1. The mission completion time increases with the number of objects. The achieved search uncertainty does not differ much in each case because the total number of cells is the same. However, the achieved classification uncertainty increases as the number of objects grows since the sensing resources get distributed.

Table 4.4 Varying number of objects.

No.	CPU	t	$E[H_S]$	$E[H_c]$
3	13.25	656.22	8.52E-06	1.16E-05
	(2.75)	(97.63)	(1.48E-06)	(2.55E-04)
5	13.33	695.53	8.13E-06	3.93E-05
	(3.29)	(113.22)	(1.94E-06)	(1.09E-04)
10	21.94	957.84	8.15E-06	4.66E-05
	(5.98)	(183.64)	(1.71E-06)	(1.01E-04)
20	36.74	1360.51	7.68E-06	1.21E-04
	(13.22)	(294.06)	(2.33E-06)	(1.67E-04)

From the above simulation results, it is concluded that the proposed algorithm is scalable for large-scale domains and a large number of objects, which is the incentive for this book.

Chapter 5
Risk-Based Sequential Decision-Making Strategy

5.1 Literature Review on Sequential Detection

The proposed approach relies on the technologies of Bayesian risk analysis. To be more specific, consider the following scenario. Imagine there are a fleet of distributed sensor-equipped MAVs with limited sensory range over a large-scale mission domain. The sensors are assumed to have measurement errors, or perception uncertainties. The goal is to detect and classify all the unknown objects within the domain with minimum risks in the presence of the noisy measurements. To achieve this objective, each vehicle sequentially updates its knowledge about object existence over the entire domain and the classification property for each found object through its own observation, which are used to compute the risks via Bayesian sequential detection method.

The key feature of sequential detection [105] is that it allows the number of observations to vary in order to achieve an optimal decision. The Bayesian sequential detection method used in this chapter is such that the Bayes risk (to be formally defined in Section 5.2.2) is minimized at each time step [133]. This method was formulated by Wald and Wolfowitz in [133] and provides a strong theoretical background for detection risk analysis. Two types of costs are taken into account in the risk calculation: 1) the cost of making a wrong decision, i.e., the probability of missed/false detection, or incorrect classification, and 2) the cost of taking more observations for a possibly better decision. The observation cost is computed in real time based on the progress of the task. Due to the randomness of observations and the dynamic observation cost, a decision may be made with a few observation samples to reduce measurement cost, whereas for other cases one would rather take more samples to reduce decision uncertainty and thus minimize the overall risk. In [148], a sequential Bayes classifier is utilized for the real-time classification of detected targets under a neural network based framework, however, without consideration of observation costs. Another sequential detection method is

Y. Wang and I.I. Hussein: Search and Classification Using MAV, LNCIS 427, pp. 89–121.
springerlink.com   © Springer-Verlag London Limited 2012

the Sequential Probability Ratio Test (SPRT) [131, 105] based on binary Neyman-Pearson formulation where no prior probability information is needed. On average, a smaller number of observations are needed to make a decision using SPRT compared with an equally reliable method with a predetermined fixed number of observations [132]. The change-point detection theory [113, 7] is a generalization and modification of SPRT. It detects a change in the probability distribution of a stochastic process or time series. Existing techniques include the Shyriaev-Roberts (SR) [113, 109] and the Cumulative Sum Control Chart (CUSUM, a.k.a. Page test) [103] tests.

In the literature, sequential decision-making via tradeoffs between exploration and exploitation has been investigated in a risk-neutral context. The work in [124, 115] and references therein provide an overview of techniques that trade off between expected information gain (or equivalently, rewards) and the cost incurred by applying a control action for Partially Observable Markov Decision Process (POMDP). The planning problem is addressed under no constraints of decision error, and is hence, risk-neutral.

5.2   Decision Making for Search and Classification

For the sake of illustration, Figure 5.1 is provided to show the block diagram of the proposed strategy and the organization of the section. At time t, the sensor takes an observation at a cell $\tilde{\mathbf{c}}_j$ in the search domain based on the sensor model proposed in Section 2.4.1. Next, the posterior probability of object existence or its classification at $\tilde{\mathbf{c}}_j$ gets updated via the Bayes update equations formulated in Sections 2.4.2 and 4.1. In Section 5.2.2, the Bayesian sequential detection method is introduced for a single cell $\tilde{\mathbf{c}}_j$, which depends on the sensor model as well as the dynamic observation cost. Its output is the minimum Bayes risk surface at cell $\tilde{\mathbf{c}}_j$. Combined with the updated probabilities, the sensor makes a decision (whether or not to take one more observation at $\tilde{\mathbf{c}}_j$) that minimizes the Bayes risk at time t. An uncertainty map is constructed based on the updated probabilities of every cell within the domain according to Sections 2.4.3 and 4.1. If the desired certainty level has not been achieved yet, a task metric is developed to formulate the dynamic observation cost. Finally, the results are combined: if the decision is to stop taking observation at the current cell $\tilde{\mathbf{c}}_j$, a sensor motion control scheme is provided, which drives the sensor to the cell $\tilde{\mathbf{c}}_k$ that has the maximum uncertainty in the domain. This process is repeated over time until both the search and classification uncertainties are satisfactorily low.

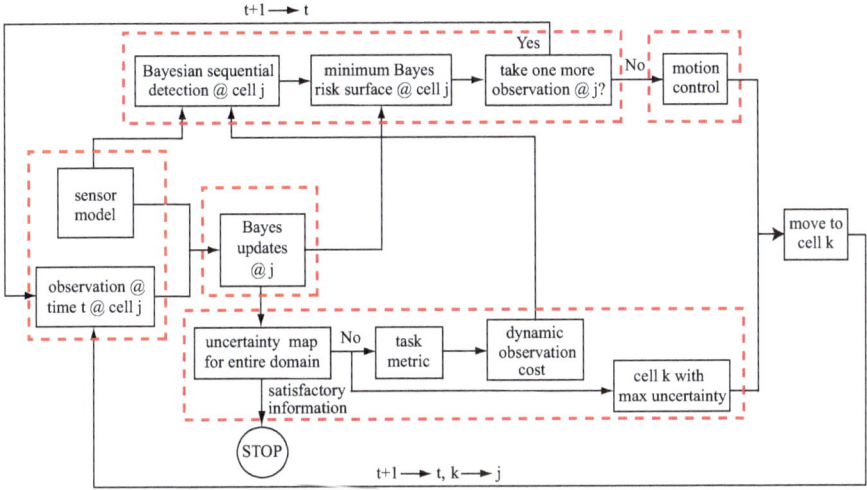

Fig. 5.1 Block diagram of cost-aware Bayesian sequential decision-making.

5.2.1 Problem Setup and Sensor Model

The sensor model proposed in Section 2.4.1 is assumed here. Conditioned on the state $X(\tilde{\mathbf{c}})$ at a particular cell $\tilde{\mathbf{c}}$, let t be time index, the observations $Y_t(\tilde{\mathbf{c}})$ taken along time are temporally i.i.d. Therefore, if a sensor takes an observation at each time step at $\tilde{\mathbf{c}}$, for a window of L time steps, there are $L+1$ different combinations of unordered scalar observations, that is, ranging from zero positive observation to L positive ones. Let the variable $Z(\tilde{\mathbf{c}})$ be the number of times that observation $Y(\tilde{\mathbf{c}}) = 1$ is taken at cell $\tilde{\mathbf{c}}$, which is a number in the set $\{0, \cdots, L\}$. The following $(L+1) \times 2$ matrix gives the general conditional probability matrix for the search task over L observations:

$$
B = \begin{bmatrix}
\text{Prob}[Z(\tilde{\mathbf{c}}) = 0 | X(\tilde{\mathbf{c}}) = 0] & \text{Prob}[Z(\tilde{\mathbf{c}}) = 0 | X(\tilde{\mathbf{c}}) = 1] \\
\text{Prob}[Z(\tilde{\mathbf{c}}) = 1 | X(\tilde{\mathbf{c}}) = 0] & \text{Prob}[Z(\tilde{\mathbf{c}}) = 1 | X(\tilde{\mathbf{c}}) = 1] \\
\vdots & \vdots \\
\text{Prob}[Z(\tilde{\mathbf{c}}) = L | X(\tilde{\mathbf{c}}) = 0] & \text{Prob}[Z(\tilde{\mathbf{c}}) = L | X(\tilde{\mathbf{c}}) = 1]
\end{bmatrix},
$$

with $\sum_{l=0}^{L} \text{Prob}[Z(\tilde{\mathbf{c}}) = l | X(\tilde{\mathbf{c}}) = j] = 1, j = 0, 1$. Because the sensor follows the Bernoulli distribution for a single observation, $\text{Prob}[Z(\tilde{\mathbf{c}}) = l | X(\tilde{\mathbf{c}}) = j]$ follows a binomial distribution with parameter β and L, which describes the probability of having l positive observations given state $X(\tilde{\mathbf{c}}) = j$. Hence, the general conditional probability matrix can be written as follows:

$$B = \begin{bmatrix} \beta^L & (1-\beta)^L \\ L(1-\beta)\beta^{L-1} & L\beta(1-\beta)^{L-1} \\ \vdots & \vdots \\ (1-\beta)^L & \beta^L \end{bmatrix}. \tag{5.1}$$

The value of β can be either the unit or the limited circular sensory range discussed in Section 2.4.1.

As stated in Section 4.1, the sensor model for the classification process follows a similar fashion with $X_c(\mathbf{p}_k)$, $Y_c(\mathbf{p}_k)$ representing the state and observation variables for object $\mathbf{p}_k, k = 1, 2, \cdots, N_0$. The general conditional probability matrix is denoted as B_c with detection probability β_c.

5.2.2 Risk-Based Sequential Decision-Making

This section takes the search process as an example to illustrate the Bayesian sequential risk analysis procedure at a single cell $\tilde{c}$. The method can be adopt to the risk-based classification of a found object $\mathbf{p}_k$ in a straightforward manner via changing parameters. Instead of deriving an optimal detector given a fixed number of observations as in classical Bayesian, Minimax or Neyman-Pearson hypothesis testing methods [105, 126, 142, 112], the Bayesian sequential detector takes observations until a decision can be made with minimum Bayes risk. This results in a random number of total observations taken.

5.2.2.1 Conditional Bayes Risk without Observation Cost

First, assuming a Uniform Cost Assignment (UCA), define the decision cost matrix as

$$C_{ij} = \begin{cases} 0 \text{ if } i = j \\ 1 \text{ if } i \neq j \end{cases},$$

where $i = 0, 1$ represent 0: object absent and 1: object present, $j = 0, 1$ correspond to state $X(\tilde{c}) = 0$ and $X(\tilde{c}) = 1$. Hence C_{ij} is the cost of deciding i when the state is $X(\tilde{c}) = j$. C can be written in the matrix form as

$$C = \begin{bmatrix} 0 & 1 \\ 1 & 0 \end{bmatrix}.$$

Let $\tilde{R}_j(\tilde{c}, L, \Delta)$, $j = 0, 1$, $L \geq 1$, be the conditional risk of deciding $X(\tilde{c}) \neq j$ at $\tilde{c}$ given that the actual state is $X(\tilde{c}) = j$ over at least one observation,

$$\tilde{R}_j(\tilde{c}, L, \Delta) = c_j \Delta b_j, \tag{5.2}$$

where

1. $c_j = [C_{0j}\ C_{1j}]$ is the jth column of the decision cost matrix C and contains the costs of deciding object absent and present given state $X(\tilde{c}) = j$.
2. $\Delta = [\Delta(i,j)]$ is the deterministic decision rule. The matrix element $\Delta(i,j)$, $i = 0, 1$, $j = 0, \cdots, L-1$ can be either 0 or 1, and $\sum_{i=0}^{1} \Delta(i,j) = 1$. When $\Delta(i,j) = 1$, it means that decision i is made given that the observation $Z = j$ corresponds to the jth column of Δ. For $L \geq 1$, the dimension of Δ is $2 \times L$ because there are two possible realizations of the states. For $L = 0$, i.e., there are no observations taken, Δ could be 'always decide there is no object', 'always decide there is an object', regardless of the observations, and there will be no explicit matrix form.
3. b_j is the jth column of the general conditional probability matrix $B = [B_{ij}]$, $i = 0, 1, \cdots, L-1$, $j = 0, 1$ for $L \geq 1$. The element B_{ij} gives the probability of having observation $Z = i$ given state j. According to the probability axiom, $\sum_{i=0}^{L-1} B_{ij} = 1$, $j = 0, 1$. For $L \geq 1$, B is a $L \times 2$ matrix.

Therefore, under UCA, there is no cost if the decision is the actual state, and the conditional risk $\tilde{R}_j$ can be interpreted as the error probability of making a wrong decision, i.e., deciding $X(\tilde{c}) \neq j$ given that the actual state is $X(\tilde{c}) = j$ under a certain decision rule Δ over L observations for cell $\tilde{c}$.

Remark 5.2.1. *"Reasonable" Deterministic Decision Rules. Here, the sensor is assumed to be a "good" one, that is to say, the detection probability is higher than the error probability of the sensor, i.e., $\beta > 0.5$. Therefore, there are only a small number of "reasonable" deterministic decision rules. Given L observations, the set of "reasonable" deterministic decision rules is the set of all rules of the type*

$$\Delta_1^l = \begin{cases} 1 & l \geq v \\ 0 & otherwise \end{cases}$$

where $l \in \{0, \ldots, L\}$ is the total number of positive observations and $v \in \{0, \ldots, L+1\}$ is the threshold where a positive decision is made. This means one only needs to consider decision rule matrices that look like

$$\Delta = \begin{bmatrix} 1 & 1 & 0 & 0 & 0 \\ 0 & 0 & 1 & 1 & 1 \end{bmatrix}$$

and not like

$$\Delta = \begin{bmatrix} 1 & 0 & 1 & 1 & 0 \\ 0 & 1 & 0 & 0 & 1 \end{bmatrix}.$$

When the threshold $v = 0$, the vehicle sensor will always decide object present and ignore the observations. Similarly, when $v = L + 1$, it will always decide object absent. Note that "reasonable" decision rules grows linearly with L and dominates any other type of decision rules with the same value of L.   •

5.2.2.2   Conditional Bayes Risk with Observation Cost

Now assign an observation cost c_{obs} each time the sensor makes a new observation. This cost could be based on energy, amount of observation time, etc. For the sake of clarity, first assume it is a constant when deriving the formulation below. A dynamic cost function $c_{obs}(t)$ is then developed to relate the observation cost with the task metrics for real-time decision-making in multi-cell domains.

Define $\phi = \{\phi_k\}_{k=0}^{\infty}$ as the stopping rule and $\delta = \{\delta_k\}_{k=0}^{\infty}$ as the intermediate decision rule. If $\phi_k = 0$, the sensor takes another measurement, if $\phi_k = 1$, the sensor stops taking further observations. At every time step k, δ_k can be either one of three intermediate decisions: (i) deciding object absent, (ii) deciding object present, or (iii) taking one more observation and postpone making a decision to the following time step. Let the stopping time be the minimum amount of time it takes to make a final decision, i.e., $N(\phi) = \min\{k : \phi_k = 1\}$, which is a random variable due to the randomness of the observations. The expected stopping time under state $X(\tilde{c}) = j$ is then given by $E_j[N(\phi)] = E[N(\phi)|X(\tilde{c}) = j]$.

Since now a cost c_{obs} is assigned for each observation, the conditional Bayes risk (5.2) under UCA over $L \geq 0$ observations can be modified as:

$$R_j(\tilde{c},L,\Delta) = \text{Prob}(\text{decide } X(\tilde{c}) \neq j | X(\tilde{c}) = j) + c_{obs}E_j[N(\phi)], \ j = 0,1. \quad (5.3)$$

If $L \geq 1$, Δ has explicit matrix form and the above equations can be rewritten as:

$$R_j(\tilde{c},L,\Delta) = c_j \Delta b_j + c_{obs}E_j[N(\phi)], \ j = 0,1. \quad (5.4)$$

5.2.2.3   Bayes Risk

Now define the Bayes risk as the expected conditional Bayes risk under decision rule Δ over L observations at cell $\tilde{c}$:

$$r(\tilde{c},L,1 - \pi_0,\Delta) = \pi_0 R_0(\tilde{c},L,\Delta) + (1 - \pi_0)R_1(\tilde{c},L,\Delta), \ L \geq 0, \quad (5.5)$$

where $\pi_0 = P(X(\tilde{c}) = 0; t = t_v)$ is the prior probability of state being $X(\tilde{c}) = 0$ at time instant t_v when an observation is taken at cell $\tilde{c}$. At each cell $\tilde{c}$ at every time step t, given a fixed π_0 under the constraints $\pi_0 \in [0,1]$, the sensor chooses a combination of $(L \geq 0, \Delta)$ that yields the minimum value of the Bayes risk r. This same procedure is repeated until the cost of making a wrong decision based on the current observation is less than that of taking one more observation for a possibly better decision.

5.2.2.4 Bayesian Sequential Detection

The following elaborates on the decision-making procedure. If the sensor does not take any observations $(L = 0)$ and directly make a decision, according to Equations (5.3) and (5.5), the Bayes risks of 2 different decision rules Δ are as follows

$$r(\tilde{c}, L = 0, 1 - \pi_0, \Delta = \text{always decide object absent}) = \pi_0,$$
$$r(\tilde{c}, L = 0, 1 - \pi_0, \Delta = \text{always decide object present}) = 1 - \pi_0.$$

If the sensor decides to take an observation $(L \geq 1)$, the minimum Bayes risk over all possible choices of Δ with L observations is

$$r_{\min}(\tilde{c}, L \geq 1, 1 - \pi_0) = \min_{\Delta \in \mathcal{G}_L} \pi_0 R_0(\tilde{c}, L \geq 1, \Delta) + (1 - \pi_0) R_1(\tilde{c}, L \geq 1, \Delta) \geq Lc_{\text{obs}}$$

where $\mathcal{G}_L$ is defined as the set of all deterministic decision rules that are based on exactly L observations.

Following similar procedure, the overall minimum Bayes risk functions $r^*_{\min}$ under all possible combinations of $(\Delta, L \geq 0)$ is computed,

$$r^*_{\min}(\tilde{c}, 1 - \pi_0) = \min_{L=0,1,2,\ldots} r_{\min}(\tilde{c}, L, 1 - \pi_0).$$

The basic procedure of Bayesian sequential detection is summarized as follows: With initial priors $\pi_j = P(X(\tilde{c}) = j; t = 0)$, $j = 0,1$, check the corresponding $r^*_{\min}$ value. If $r^*_{\min}$ is given by the risk function with $L \geq 1$, the sensor takes an observation $Y_{t=0}(\tilde{c})$. Compute the posteriors $P(X(\tilde{c}) = j | Y_{t=0}(\tilde{c}); t = 1)$ according to Equation (2.33) and again check $r^*_{\min}$ to make decisions. The process is repeated using these posteriors as the new priors. The key is that an observation is taken if and only if $r_{\min}(\tilde{c}, L \geq 1, 1 - \pi_0) < \min(1 - \pi_0, \pi_0)$. When $r^*_{\min} = r_{\min}(\tilde{c}, L = 0, 1 - \pi_0)$, the sensor stops taking observations and a decision is made at $\tilde{c}$.

5.2.2.5 Simulation for a Single Cell

The following preliminary simulation for a single cell illustrates the proposed scheme. Fix a cell $\tilde{c}$, choose $\beta = 0.8$ (i.e., $M = 0.3$ and the sensor is right located at the centroid of this cell), and set the observation cost as a fixed number $c_{\text{obs}} = 0.05$ to demonstrate the Bayesian sequential detection method. Figure 5.2(a) shows all the Bayes risk functions r under 0 (black lines), 1 (blue lines) and 2 (green lines) observations with $\pi_0 \in [0,1]$. In Figure 5.2(b), the red segment indicates the overall minimum Bayes risk $r^*_{min}(\tilde{c}, 1 - \pi_0)$. The overall minimum Bayes risk curve $r^*_{min}(\tilde{c}, 1 - \pi_0)$ is constructed by taking the smallest value of all $r_{min}(\tilde{c}, L, 1 - \pi_0), L = 0, 1, 2, \cdots$ under each fixed prior probability π_0. Figure 5.2(c) shows the construction of the minimum Bayes risk (the red dot) under a fixed prior π_0^*. Here, only the lines of decision rules that constitute the red segment are shown with the corresponding equations listed. The Bayes risk functions under more than 3 observations ($L \geq 3$) have larger r values and do not contribute to $r^*_{min}(\tilde{c}, 1 - \pi_0)$ for the particular choice of β and c_{obs} here.

Each of the lines is interpreted as follows.

Line 1. This line represents the decision rules without any observation. Always decide there is an object at the cell regardless of the observations. According to Equation (5.5),

$$r(\tilde{c}, L = 0, 1 - \pi_0, \Delta = \text{always decide there is an object})$$
$$= \pi_0 \times 1 + (1 - \pi_0) \times 0 = \pi_0;$$

Line 2. This line also represents the decision rules without any observation. Always decide there is no object regardless of the observations:

$$r(\tilde{c}, L = 0, 1 - \pi_0, \Delta = \text{always decide there is no object})$$
$$= \pi_0(0 + c_{obs} \times 0) + (1 - \pi_0)(1 + c_{obs} \times 0) = 1 - \pi_0.$$

Line 3. The blue line corresponds to the decision rule 3 after taking one observation: decide the actual state according to the only one observation, that is, if $Z = 1$, decide there is actually an object. It follows that

$$r(\tilde{c}, L = 1, 1 - \pi_0, \Delta = \Delta_{11})$$
$$= \pi_0(1 - \beta + c_{obs}) + (1 - \pi_0)(1 - \beta + c_{obs}) = 1 - \beta + c_{obs}.$$

Line 4. This line gives the decision rules after two observations. Line 4 corresponds to the decision rule that decides there is actually an object if and only if all the two observations are positive ($Z = 2$). Following the same procedure as above, it follows that

$$r(\tilde{c}, L = 2, 1 - \pi_0, \Delta = \Delta_{21})$$
$$= (1 - \beta)^2 \pi_0 + (2\beta(1 - \beta) + (1 - \beta)^2)(1 - \pi_0) + 2c_{obs};$$

Line 5. This line also gives the decision rules after two observations. Line 5 corresponds to the decision rule that decides there is no object if and only if none of the two observations is object present,

$$r(\tilde{c}, L = 2, 1 - \pi_0, \Delta = \Delta_{22})$$
$$= (2\beta(1 - \beta) + (1 - \beta)^2)\pi_0 + (1 - \beta)^2(1 - \pi_0) + 2c_{obs}.$$

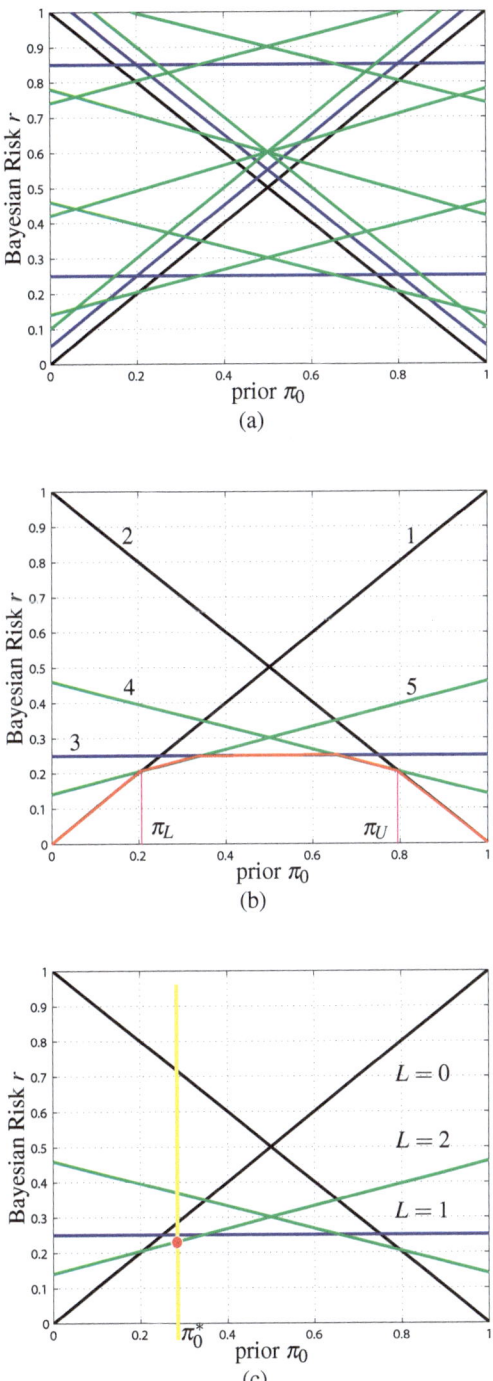

Fig. 5.2 Bayes risk, minimum Bayes risk, and construction of minimum Bayes risk.

Thus, the red segment gives the minimum Bayesian risk $r^*_{\min}(\tilde{\mathbf{c}}, 1 - \pi_0)$ over 0,1,2 observations.

The intersection of lines 1, 5 is the lower prior probability $\pi_L = 0.2059$. When the posterior probability $P(X(\tilde{\mathbf{c}}) = 0, t)$ updated through Equation (2.33) is below π_L, the vehicle sensor stops taking observation and decides that the actual state is object present. This is because the minimum Bayesian risk is determined by line 1 instead of line 5 when $P(X(\tilde{\mathbf{c}}) = 0, t) \in [0, \pi_L]$. The intersection of lines 2, 4 is the upper prior probability $\pi_U = 0.7941$. When $P(X(\tilde{\mathbf{c}}) = 0, t)$ is above π_U (i.e., $P(X(\tilde{\mathbf{c}}) = 1, t) \le \pi_L$), the sensor decides that there is actually no object.

The following simple example illustrates how to utilize the minimum Bayes risk curve $r^*_{\min}$ for decision-making. At a cell $\tilde{\mathbf{c}}$, assume the initial prior $P(X(\tilde{\mathbf{c}}) = 0, t) = P(X(\tilde{\mathbf{c}}) = 1, t) = 0.5$. The corresponding minimum Bayes risk for the prior 0.5 is given by Line 3. So the sensor takes one observation, and if the observation is $Y_{t=1}(\tilde{\mathbf{c}}) = 1$ indicating there is an object, the posterior probability is updated according to the new observation and the Bayes update rules (2.33). The posterior probability is $P(X(\tilde{\mathbf{c}}) = 1, t) = 0.8, P(X(\tilde{\mathbf{c}}) = 0, t) = 0.2 \le \pi_L$. Now $r^*_{\min}$ is given by Line 1. Therefore, the sensor decides not to take any more observation and determine there is actually an object at this cell with Bayes risk $r = 0.2$.

5.2.3 Extension to Full-Scale Domain

The mechanics of the Bayesian probability updates (Section 2.4.2) and Bayesian sequential detection (Section 5.2.2) have been discussed for a single cell. This section defines an uncertainty map based on these posterior probabilities and the metrics for the search and classification tasks in general multi-cell domains. The search task metric is related with a dynamic observation cost for the Bayesian sequential decision-making strategy in multi-cell domains. Based on these, the sensor motion control laws in Section 2.4.4 is used, which seeks to find and classify all objects in $\mathscr{D}$ with a desired confidence level.

As stated in Sections 2.4.2 and 4.1, Bayes rule is used to update 1) the probability of object present at each cell $\tilde{\mathbf{c}}$ in $\mathscr{D}$ and 2) the probability of object having Property 'G' for each found object $\mathbf{p}_k$. These updated posterior probabilities are then used to construct the uncertainty functions (2.35) and (4.2) for the search and classification process, respectively.

The search and classification metrics (4.2) and (4.3) developed in Section 4.2 are used here for the risk-based decision-making for search versus classification. Define the classification conditions as follows:

$$\begin{cases} \|\mathbf{q}(t) - \mathbf{p}_k\| \le r_c & \text{(a)} \\ H_c(P_{H_c}, \mathbf{p}_k, t) > H_d(\mathbf{p}_k, t) & \text{(b)} \\ H_s(P_{H_s}, \mathbf{p}_k, t) \le H_s^U & \text{(c)} \\ \text{No Decision at } \mathbf{p}_k \text{ at } t & \text{(d)} \end{cases}, \qquad (5.6)$$

where H_s^{U} is some upper bound on the search uncertainty to be met before a classification task can be carried on. Only when all the classification conditions are satisfied, i.e., (a) the object $\mathscr{O}_k$ is within the vehicle's classification sensory range, (b) the classification uncertainty of $\mathscr{O}_k$ is larger than the desired uncertainty, (c) the search uncertainty of $\mathscr{O}_k$ is relatively low (It is to some extent sure that $\mathscr{O}_k$ is an object), and (d) no decision has been made about the property of $\mathscr{O}_k$ yet at previous time step, then the vehicle will start to classify $\mathscr{O}_k$. If any one of the above condition fails, the vehicle $\mathscr{V}$ stop classifying the found object and switch to searching again. It can resume classifying an object that has been detected and completely or partially classified in the past if it finds it again during the search process. When this occurs, the value of H_d will be smaller than the last time the objected has been detected.

Now associate a dynamic observation cost $c_{\mathrm{obs}}(t)$ with the search cost function $\mathscr{J}(t)$,

$$c_{\mathrm{obs}}(t) = \gamma \mathscr{J}(t), \qquad (5.7)$$

where $\gamma > 0$ is some positive weighting parameter. At the outset of the mission, few regions in the domain have been covered, therefore, the cost, $\mathscr{J}$, of not searching anywhere else is high. Equivalently, taking an observation at the current cell is "expensive", i.e., $c_{\mathrm{obs}}(t)$ is large. In this case, the risk-based sequential decision-making strategy tends to make a decision with a few observations, which may yield large number of wrong decisions (however, it still gives the minimum Bayes risk over all decisions given the limited available observations), but increase the potential of rapidly detecting and classifying more critical objects in the domain. When the sensor stops taking observations, makes a decision, and leaves the current cell, it will move to another cell and again take an observation there. Because the uncertainty level associated with that cell changes (Equations (2.33),(2.35)), the values for $\mathscr{J}$ (Equation (4.2)) and c_{obs} (Equation (5.7)) over the entire domain differ accordingly. Additional information is gained by changing the cell to be observed. When the sensor has surveyed more regions in the domain, the uncertainty level at all the visited cells is reduced with respect to the initial uncertainty, and hence both $\mathscr{J}$ and c_{obs} decrease. The process will be repeated until $\mathscr{J}(t) \to 0$ and $H_c \to 0$, $\forall \mathbf{p}_k$, i.e., all the unknown objects of interest within the domain have been found and classified with a desired uncertainty level in a small neighborhood of zero. Note that the observation cost is assigned according to the real-time progress of the search and classification tasks and facilitates real-time decision-making based on the available observations.

Remark 5.2.2. *A small value of γ corresponds to the case where the sensor will stay in a cell until a high certainty about object existence or its classification is achieved before moving on. A large value gives the opposite case, i.e., the sensor will not linger long in any cell until it has had a chance to survey more regions in the domain.* ●

5.2.4 Simulation

In this simulation, consider all the cells $\tilde{\mathbf{c}}$ within a 20×20 square domain $\mathscr{D}$. For each $\tilde{\mathbf{c}} \in \mathscr{D}$, an i.i.d. prior probability of object presence is assumed, which equals to $P(X(\tilde{\mathbf{c}}) = 1, 0) = \frac{E[N_0]}{N_{\text{tot}}} = 0.2$, where $E[N_0] = 80$ is the expected number of objects. For the classification process, let the desired upper bound for classification uncertainty be $H_c^u = 0.01$ and $H_s^U = 0.3$. The priors $P_c(X_c(\mathbf{p}_k) = 0, 0) = 0.5$, $\forall k$ and all the objects with even number have property 'G'. The locations of the objects are randomly generated. The number of objects generated for this simulation turns out to be 83. The locations of objects with Property 'F' are indicated by the 42 green crosses and the locations of the objects with Property 'G' are indicated by the 41 magenta crosses in Figure 5.3. Figure 5.3 shows the evolution of the search uncertainty map H_s (dark red for highest uncertainty and dark blue for lowest uncertainty) at (a) $t = 1$, (b) $t = 200$, (c) $t = 400$, and (d) $t = 800$. The radius r of the search sensor is chosen to be 8 and the classification radius r_c is chosen to be 6, as shown by the magenta and green circles in Figure 5.3. Set the maximum sensing capacity as $M = 0.5$. The parameter $\gamma = 0.05$. The black dot represents the position of the vehicle. Here the control law in Equation (2.36) is used with control gain $\bar{\bar{k}} = 0.2$. The set $\mathscr{U}$ is chosen to be $\mathscr{D}$. From the simulation results, it can be concluded that at most $H_s = 1.1 \times 10^{-6}$ has been achieved everywhere within $\mathscr{D}$.

Figure 5.4(a) records the number of false detections and missed detections versus time. It can be seen from the figure that the number of missed detections (18) is much larger than that of the false detections (2) at the beginning of the task. This is because the initial prior probability $P(X(\tilde{\mathbf{q}}) = 1, 0)$ to start with is closer to zero, which makes it easier to make a wrong decision after taking an erroneous observation $Y(\tilde{\mathbf{q}}) = 0$ given that the actual state is object present. Figure 5.4(b) compares the number of incorrect classifications, i.e., deciding Property 'F' given Property 'G', and deciding Property 'G' given Property 'F' over all detected objects. These two numbers are similar since $P_c(X_c(\tilde{\mathbf{c}}), 0) = 0.5$. In both figures, it can be shown that as time increases, the number of missed detections and false detections decrease. Both of the error numbers go to zero with zero uncertainty at the end of the mission. This implies that one can balance between the number of errors within the tolerance range and the limited time to decide when to stop.

Figure 5.5(a) shows the classification results for object 1. Its probability of having Property 'G' is zero and the corresponding uncertainty function $H_c = 0$, i.e., it is 100% sure that object 1 has Property 'F'. Similarly, Figure 5.5(b) shows that object 2 has Property 'G' with zero uncertainty. The properties of other objects are also satisfied classified with the desired uncertainty level and can be shown like Figures 5.5(a) and 5.5(b) without difficulty.

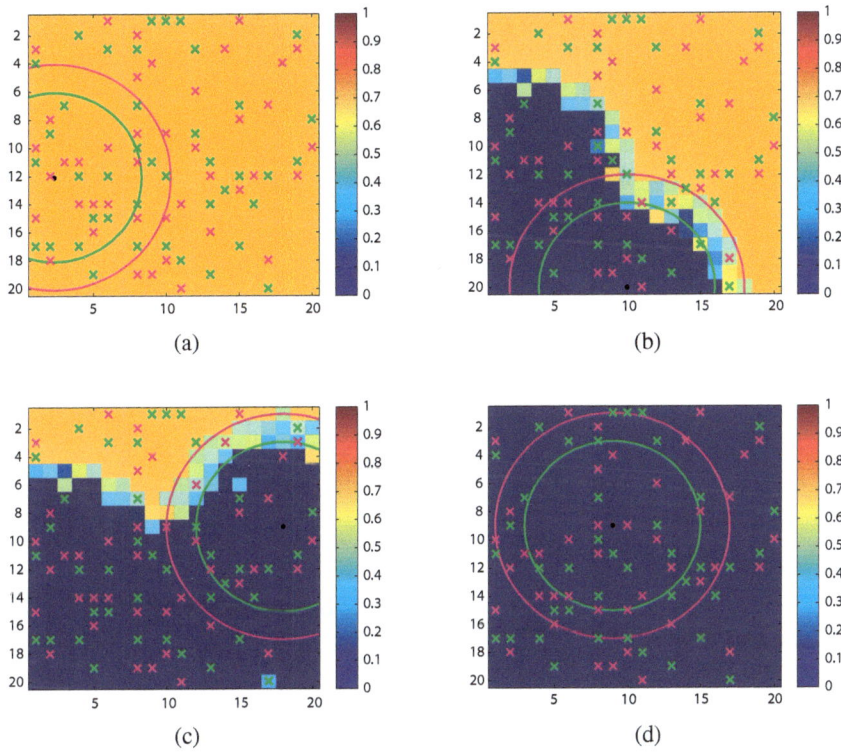

Fig. 5.3 Evolution of search uncertainty.

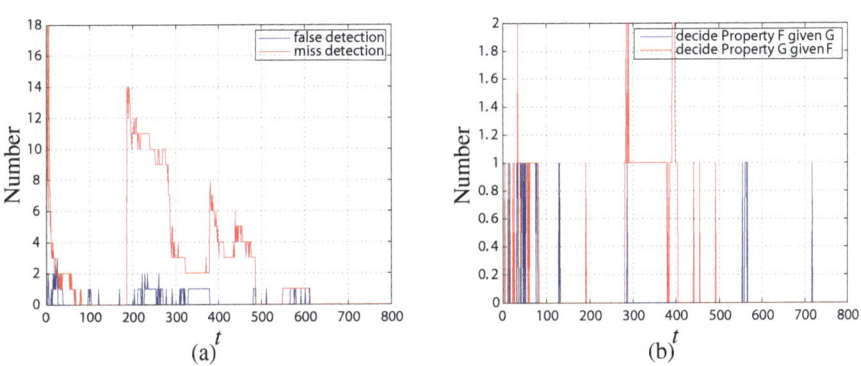

Fig. 5.4 Number of false/missed detections, and incorrect classifications.

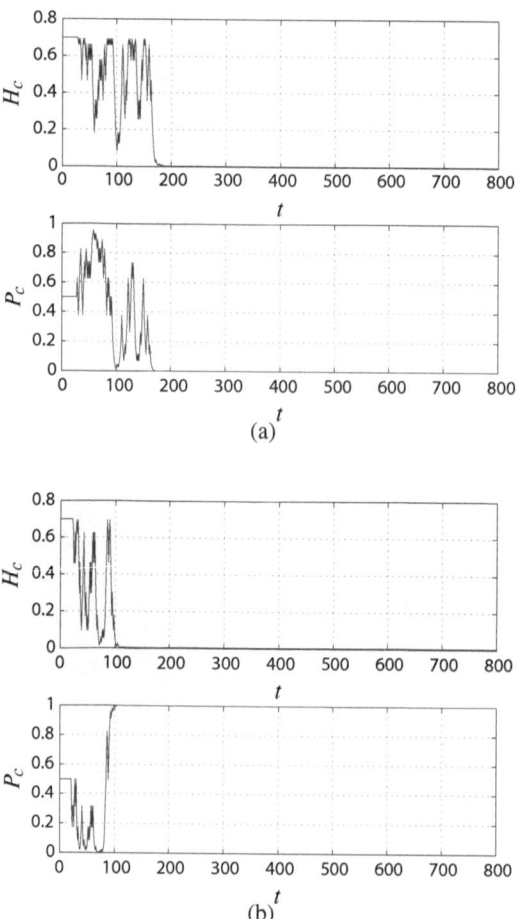

Fig. 5.5 Classification results for objects 1 and 2.

5.3 Extension to Three States

In this section, the above standard binary Bayesian sequential detection method is extended into a ternary risk-based sequential decision-making strategy. This allows concurrent search and classification observations taken by a single autonomous vehicle sensor. However, the decision to be made here is still the same, i.e., whether to make a prompt decision regarding object existence or its classification based on insufficient observations, or to keep taking observations at the current location until 100% certain about the true state.

5.3.1 Problem Setup and Sensor Model

Now let $X(\tilde{c})$ be a ternary state random variable at cell $\tilde{c}$, where 0 corresponds to object absent, 1 corresponds to object having Property 'F', and '2' corresponds to object having Property 'G'.

For the sake of illustrative clarity, the following assumptions for the sensor model are made.

1. A sensor is able to observe only one cell at a time. That is, the unit-range sensor model is assumed in this section. Extension to other sensor models that are capable of observing multiple cells at the same time (e.g., the sensor models with limited sensory range proposed in [61, 59, 60, 135, 62, 136, 140, 138, 137, 139]) is straightforward.
2. A sensor is able to move to any cell within the domain. Other motion schemes, such as gradient-based, awareness-based, and information-driven control laws ([61, 59, 60, 135, 62, 136, 140, 138, 137, 139]) can be adopted without difficulty.

Let $Y(\tilde{c})$ be the corresponding ternary observation random variable. The sensor model follows a ternary discrete probability distribution. For a cell $\tilde{c}$, given a state $X(\tilde{c}) = i$, $i = 0, 1, 2$, the probability mass function f of the observation distribution is given by

$$f_Y(y|X(\tilde{c}) = i) = \begin{cases} \beta_{i0} & \text{if } y = 0 \\ \beta_{i1} & \text{if } y = 1 \\ \beta_{i2} & \text{if } y = 2 \end{cases}, \tag{5.8}$$

where $\sum_{j=0}^{2} \beta_{ij} = 1$, Y corresponds to the ternary random variable and y is the dummy variable. Figure 5.6 shows the relationship between the unknown state $X(\tilde{c})$ and an observation $Y(\tilde{c})$.

Conditioned on the true state $X(\tilde{c})$, let t be the time index, the observations $Y_t(\tilde{c})$ taken along time are temporally i.i.d. Define an integer random variable $Z_j(\tilde{c})$, $j = 0, 1, 2$ as the number of times that observation $Y(\tilde{c}) = j$ appears during a window of L time steps. The quantity $Z_j(\tilde{c})$ satisfies $\sum_{j=0}^{2} Z_j(\tilde{c}) = L$, $Z_j(\tilde{c}) \in [0, L]$. Therefore, given state $X(\tilde{c}) = i$, $i = 0, 1, 2$, the probability of having observation (z_0, z_1, z_2) in a window of L time steps follows a multinomial distribution

$$\text{Prob}(Z_0(\tilde{c}) = z_0, Z_1(\tilde{c}) = z_1, Z_2(\tilde{c}) = z_2|X(\tilde{c}) = i) = \frac{L!}{z_0! z_1! z_2!} \beta_{i0}^{z_0} \beta_{i1}^{z_1} \beta_{i2}^{z_2}. \tag{5.9}$$

The sensor's probabilities of making a correct observation, i.e., the detection probabilities, are β_{00}, β_{11} and β_{22}. Here it is assumed that the sensor is "good" and restrict these values to be $\beta_{00}, \beta_{11}, \beta_{22} > 0.5$. More general values within $[0, 1]$ can be considered, however, introducing extra analytical complexity that does not contribute any new insights. It is assumed that the sensor's probabilities of making an

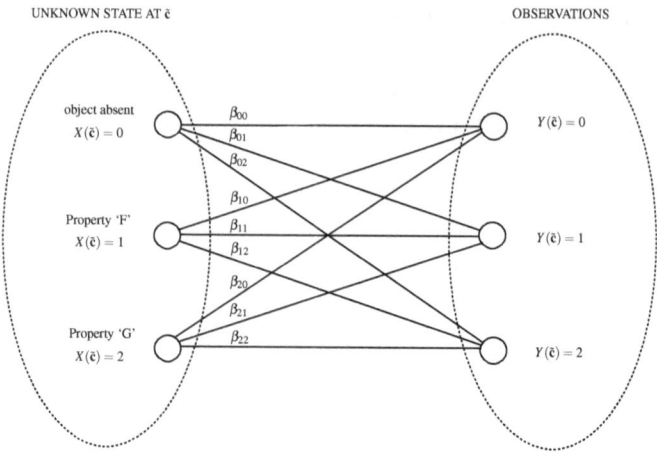

Fig. 5.6 Ternary sensor model.

erroneous observation, i.e., the error probabilities, β_{ij}, $i \neq j$, follow a simple linear model under the probability axiom constraint $\sum_{j=0}^{2} \beta_{ij} = 1$:

$$\beta_{ij} = v_j(1 - \beta_{ii}), \quad i \neq j, \tag{5.10}$$

where v_j is some weighting parameter that satisfies $\sum_{j \neq i} v_j = 1$, $0 \leq v_j \leq 1$. This implies that the sensor is able to better distinguish the true state from the other two states and returns an higher likely observation of the true state at that location.

5.3.2 Ternary Bayesian Updates for Search and Classification

According to Bayes' rule, given a single observation $Y_t(\tilde{c}) = j$ taken at cell $\tilde{c}$ at time step t, it follows that

$$\begin{aligned} &P(X(\tilde{c}) = i | Y_t(\tilde{c}) = j; t+1) \\ &= \eta_j P(Y_t(\tilde{c}) = j | X(\tilde{c}) = i) P(X(\tilde{c}) = i; t), \quad i, j = 0, 1, 2. \end{aligned} \tag{5.11}$$

where $P(Y_t(\tilde{c}) = j | X(\tilde{c}) = i)$ is determined by the ternary sensor model (5.8), and the β_{ii} and β_{ij} $(i \neq j)$ function (5.10).

According to the law of total probability, η_j is given as follows,

$$\eta_j = \frac{1}{P(Y_t(\tilde{c}) = j)} = \frac{1}{\beta_{0j}P(X(\tilde{c}) = 0; t) + \beta_{1j}P(X(\tilde{c}) = 1; t) + \beta_{2j}P(X(\tilde{c}) = 2; t)},$$

and thus the posterior probabilities is given by substituting the value of η_j into Equation (5.11).

5.3.3 Ternary Risk-Based Sequential Decision-Making

In this section, a ternary risk-based sequential decision-making strategy is used to determine the state at a cell $\tilde{c}$ with minimum Bayes risk. It is extended from the above standard binary Bayesian sequential detection method [105, 106, 132] in signal detection theory [105, 126, 142, 112]. The formulation for Bayes risk in ternary case is similar as the binary case, however, it ends up with the minimum Bayes risk surface instead of minimum Bayes risk curve. Here only the main results are listed and a simulation at a single cell is used to illustrate the modified methods.

The ternary conditional Bayes risk under UCA over $L \geq 0$ observations is as follows:

$$R_j(\tilde{c}, L, \Delta) = \text{Prob}(\text{decide } X(\tilde{c}) \neq j | X(\tilde{c}) = j) + c_{\text{obs}} E_j[N(\phi)], \quad j = 0, 1, 2. \ (5.12)$$

If $L \geq 1$, Δ has explicit matrix form and the above equations can be rewritten as:

$$R_j(\tilde{c}, L, \Delta) = c_j \Delta b_j + c_{\text{obs}} E_j[N(\phi)], \quad j = 0, 1, 2. \tag{5.13}$$

where

1. $c_j = [C_{0j} \ C_{1j} \ C_{2j}]$.
2. $\Delta = [\Delta(i, n)]$ is the deterministic decision rule. Let $\mathcal{N}$ be the total number of possible observation combinations (z_0, z_1, z_2) that the sensor can take according to the multinomial distribution (5.9) over a window of L time steps. For $L \geq 1$, the dimension of Δ is $3 \times \mathcal{N}$. For $L = 0$, i.e., there are no observations taken, Δ could be 'always decide there is no object', 'always decide there is an object with Property 'F'' or 'always decide there is an object with Property 'G''.
3. b_j is the jth column of $B = [B_{ij}]$ for $L \geq 1$. For $L \geq 1$, B is a $N \times 3$ matrix.

The ternary Bayes risk for $L \geq 0$ is given as follows

$$r(\tilde{c}, L, \pi_1, \pi_2, \Delta) = (1 - \pi_1 - \pi_2) R_0(\tilde{c}, L, \Delta) + \pi_1 R_1(\tilde{c}, L, \Delta) + \pi_2 R_2(\tilde{c}, L, \Delta), \ (5.14)$$

where $\pi_j = P(X(\tilde{c}) = j; t = t_v)$, $j = 0, 1, 2$ is the prior probability of state being $X(\tilde{c}) = j$ at time instant t_v when an observation is taken at cell $\tilde{c}$.

If the sensor does not take any observations ($L = 0$) and directly makes a decision, the Bayes risks of 3 different decision rules Δ are as follows

$$r(\tilde{c}, L = 0, \pi_1, \pi_2, \Delta = \text{always decide object absent}) = \pi_1 + \pi_2,$$
$$r(\tilde{c}, L = 0, \pi_1, \pi_2, \Delta = \text{always decide object having Property 'F'}) = 1 - \pi_1,$$
$$r(\tilde{c}, L = 0, \pi_1, \pi_2, \Delta = \text{always decide object having Property 'G'}) = 1 - \pi_2.$$

The overall minimum Bayes risk over all possible combinations of (Δ, L) is,

$$r_{\min}^*(\tilde{c}, \pi_1, \pi_2) = \min_{L = 0, 1, 2, \ldots, \Delta \in \mathscr{G}_L} r(\tilde{c}, L, \pi_1, \pi_2, \Delta).$$

An observation is taken if and only if $\min_{\Delta \in \mathscr{G}_L} r(\tilde{c}, L \geq 1, \pi_1, \pi_2, \Delta) < \min(\pi_1 + \pi_2, 1 - \pi_1, 1 - \pi_2)$.

The following preliminary simulation for a single cell is used to illustrate the proposed scheme. Fix a cell $\tilde{c}$ and assume that the sensor is located at the centroid of this cell. The sensing parameters are chosen as follows:

$$\begin{aligned} \beta_{00} &= 0.8, \beta_{01} = 0.1, \beta_{02} = 0.1, \\ \beta_{10} &= 0.2, \beta_{11} = 0.7, \beta_{12} = 0.1, \\ \beta_{20} &= 0.1, \beta_{21} = 0.15, \beta_{22} = 0.75. \end{aligned} \qquad (5.15)$$

Figure 5.7(a) shows all the Bayes risk functions r under $L = 0$, 1 or 2 observations under the constraints $\pi_i \in [0, 1]$ and $\sum_{i=1}^{2} \pi_i \leq 1$. Figure 5.7(b) shows the overall minimum Bayes risk surface $r^*_{\min}(\tilde{c}, \pi_1, \pi_2)$, which is the minimum value of all $r(\tilde{c}, L, \pi_1, \pi_2, \Delta)$, $L \geq 0$, under each fixed prior probability pair (π_1, π_2). The overall minimum risk surface is composed of several enumerated risk planes, each of which is described briefly in this section.

Each of these risk planes in Figure 5.7(b) annotated by the numerals $1 - 10$ is interpret as follows.

Risk Plane 1. $r(\tilde{c}, L = 0, \pi_1, \pi_2, \Delta = \text{always decide there is no object}) = \pi_1 + \pi_2$.
Risk Plane 2. $r(\tilde{c}, L = 0, \pi_1, \pi_2, \Delta = \text{always decide object present with Property 'F'})$ $= 1 - \pi_1$.
Risk Plane 3. $r(\tilde{c}, L = 0, \pi_1, \pi_2, \Delta = \text{always decide object present with Property 'G'})$ $= 1 - \pi_2$.
Risk Plane 4. This plane corresponds to the decision rule after taking one observation. The general conditional probability matrix for $L = 1$ is given as

$$B(L = 1) = \begin{bmatrix} \beta_{00} & \beta_{10} & \beta_{20} \\ \beta_{01} & \beta_{11} & \beta_{21} \\ \beta_{02} & \beta_{12} & \beta_{22} \end{bmatrix},$$

where the rows correspond to the observations $(z_0 = 1, z_1 = 0, z_2 = 0)$, $(z_0 = 0, z_1 = 1, z_2 = 0)$, and $(z_0 = 0, z_1 = 0, z_2 = 1)$, respectively. Risk Plane 4 corresponds to the following decision rule,

$$\Delta_{11} = \begin{bmatrix} 1 & 0 & 0 \\ 0 & 1 & 0 \\ 0 & 0 & 1 \end{bmatrix}.$$

That is, decide the state according to the only one observation taken. This is the only reasonable decision rule for $L = 1$. Therefore, according to Equation (5.13), it follows that $R_0(\tilde{c}, L = 1, \Delta = \Delta_{11}) = \beta_{01} + \beta_{02} + c_{\text{obs}}$, $R_1(\tilde{c}, L = 1, \Delta = \Delta_{11}) = \beta_{10} + \beta_{12} + c_{\text{obs}}$, and $R_2(\tilde{c}, L = 1, \Delta = \Delta_{11}) = \beta_{20} + \beta_{21} + c_{\text{obs}}$. Hence, $r(\tilde{c}, L = 1, \pi_1, \pi_2, \Delta = \Delta_{11})$ is given directly by Equation (5.14).

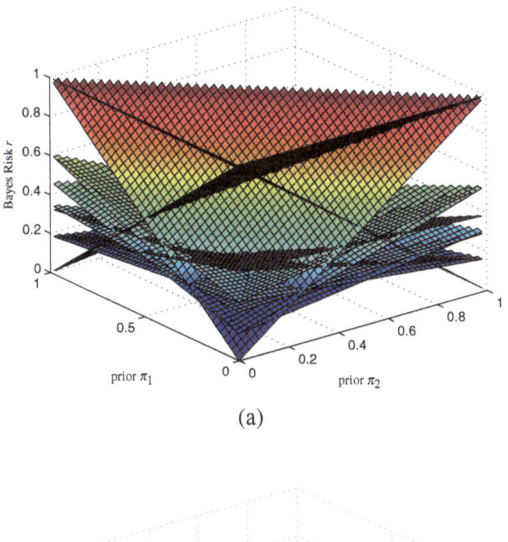

(a)

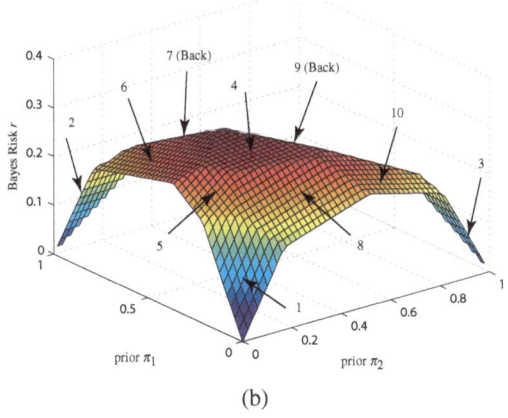

(b)

Fig. 5.7 Bayes risk surface and minimum Bayes risk surface.

Risk Planes 5-10. These plane give the decision rules after two observations. The general conditional probability matrix for $L = 2$ is given as

$$B(L=2) = \begin{bmatrix} \beta_{00}^2 & \beta_{10}^2 & \beta_{20}^2 \\ \beta_{01}^2 & \beta_{11}^2 & \beta_{21}^2 \\ \beta_{02}^2 & \beta_{12}^2 & \beta_{22}^2 \\ 2\beta_{00}\beta_{01} & 2\beta_{10}\beta_{11} & 2\beta_{20}\beta_{21} \\ 2\beta_{00}\beta_{02} & 2\beta_{10}\beta_{12} & 2\beta_{20}\beta_{22} \\ 2\beta_{01}\beta_{02} & 2\beta_{11}\beta_{12} & 2\beta_{21}\beta_{22} \end{bmatrix},$$

where the rows correspond to the observations $(z_0 = 2, z_1 = 0, z_2 = 0)$, $(z_0 = 0, z_1 = 2, z_2 = 0)$, $(z_0 = 0, z_1 = 0, z_2 = 2)$, $(z_0 = 1, z_1 = 1, z_2 = 0)$, $(z_0 = 1, z_1 = 0, z_2 = 1)$, and $(z_0 = 0, z_1 = 1, z_2 = 1)$, respectively. The corresponding decision rules are,

$$\Delta_{21} = \begin{bmatrix} 1\,0\,0\,1\,1\,0 \\ 0\,1\,0\,0\,0\,1 \\ 0\,0\,1\,0\,0\,0 \end{bmatrix} (5), \quad \Delta_{22} = \begin{bmatrix} 1\,0\,0\,0\,1\,0 \\ 0\,1\,0\,1\,0\,1 \\ 0\,0\,1\,0\,0\,0 \end{bmatrix} (6)$$

$$\Delta_{23} = \begin{bmatrix} 1\,0\,0\,0\,0\,0 \\ 0\,1\,0\,1\,0\,1 \\ 0\,0\,1\,0\,1\,0 \end{bmatrix} (7), \quad \Delta_{24} = \begin{bmatrix} 1\,0\,0\,1\,1\,0 \\ 0\,1\,0\,0\,0\,0 \\ 0\,0\,1\,0\,0\,1 \end{bmatrix} (8)$$

$$\Delta_{25} = \begin{bmatrix} 1\,0\,0\,0\,0\,0 \\ 0\,1\,0\,1\,0\,0 \\ 0\,0\,1\,0\,1\,1 \end{bmatrix} (9), \quad \Delta_{26} = \begin{bmatrix} 1\,0\,0\,1\,0\,0 \\ 0\,1\,0\,0\,0\,0 \\ 0\,0\,1\,0\,1\,1 \end{bmatrix} (10)$$

The Bayes risks follow according to Equations (5.13) and (5.14).

When r^*_{min} is given by Risk Plane 1, 2 or 3, the sensor stops taking observation and makes the final decision, otherwise, it always takes one more observation.

5.3.4   The Uncertainty Map, Task Metric, and Motion Control

Let P_H be the probability distribution for object absent and its classification at cell $\tilde{c}$ at time t and is given by $P_H = \{P(X(\tilde{c}) = 0;t), P(X(\tilde{c}) = 1;t), P(X(\tilde{c}) = 2;t)\}$. Define its information entropy as:

$$H(P_H, \tilde{c}, t) = -\sum_{j=0}^{2} P(X(\tilde{c}) = j;t) \ln P(X(\tilde{c}) = j;t). \qquad (5.16)$$

The maximum value attainable by H is $H_{max} = 1.0986$ when $P(X(\tilde{c}) = j, t) = \frac{1}{3}$.

Define the associated cost of not carrying further search and classification as follows:

$$\mathscr{J}(t) = \frac{\sum_{\tilde{c} \in \mathscr{D}} H(P_H, \tilde{c}, t)}{H_{max} A_{\mathscr{D}}}. \qquad (5.17)$$

A dynamic observation cost $c_{obs}(t)$ is assumed according to Equation (5.7).

Next, consider a control strategy for the sensor motion over the mission domain $\mathscr{D}$. Combining with the Bayesian sequential decision-making strategy, it seeks to find and classify all objects in $\mathscr{D}$ with a desired confidence level (i.e., achieve $\mathscr{J} \to 0$) under a dynamic observation cost and the minimum Bayes risk at every time step. As mentioned in Section 5.3.1, it is assumed that there is no speed limit on the sensor, i.e., the sensor is able to move to any cell within $\mathscr{D}$ from its current location.

The memoryless motion control scheme presented in Section 2.4.5.4 is adopted here, where the sets $\mathscr{Q}_H(t)$ and $\mathscr{Q}_d(t)$ are obtained by considering a single autonomous vehicle sensor in Equations (2.48, 2.49).

5.3.5   Full-Scale Domain Simulations

This section provides A) a detailed numerical simulation that illustrates the performance of the decision-making strategy, and B) a Monte-Carlo simulation comparison between the proposed strategy and the classical fixed-sample Bayesian hypothesis testings. All the simulations are implemented on a 2.80-GHz, i7-860m processor with 4.0GB RAM, and Matlab®-compiled codes.

5.3.5.1   Simulation Example

Consider a 20×20 square domain $\mathscr{D}$. For each $\tilde{\mathbf{c}} \in \mathscr{D}$, assume an i.i.d. prior probability distribution: $P(X(\tilde{\mathbf{c}}) = 0; t = 0) = 0.7$, $P(X(\tilde{\mathbf{c}}) = 1; t = 0) = 0.1$, and $P(X(\tilde{\mathbf{c}}) = 2; t = 0) = 0.2$. The sensing parameters β_{ij} are the same as in Equation (5.15). The observation cost weighting parameter γ in Equation (5.7) is set as 0.05 and the desired uncertainty for every cell is 0.02.

The number of objects generated for this simulation turns out to be 125 (the expected number of objects is 120 according to Equation (2.29)) with 64 objects with Property 'F' and 61 objects with Property 'G'.

Table 5.1 shows the mean percentage of missed detections, false detections, and incorrect classifications during time period $1 - 200, 201 - 400, 401 - 600$, and $601 - 800$, respectively. 100 runs are carried out in 800 time steps with the same parameter settings as above. From the table, most of the errors occur at the earlier stage of the mission and the number of errors decreases with time.

Table 5.1 Mean percentage of wrong decisions during different time periods.

mean percentage (%)	$1 - 200$	$201 - 400$	$401 - 600$	$601 - 800$
missed detection	40.43	34.51	17.35	7.71
false detection	43.49	25.43	21.64	9.44
incorrect classification	42.09	24.48	18.52	14.91

5.3.5.2   Monte-Carlo Simulation Comparison

Now a Monte-Carlo simulation is performed to compare the performance of the proposed Bayesian sequential strategy and the classical fixed-sample Bayesian hypothesis testing [105, 25, 112]. Under UCA, the fixed-sample Bayesian hypothesis testing is the maximum a posterior (MAP) estimator. That is, the optimal decision corresponds to the state that gives the maximum posterior probability after L observations. Note that this is an off-line batch technique where a decision is made if and only if all the fixed L observations have been taken. Here it is used as a benchmark performance criterion.

From the simulation results, the expected number of observations taken at each cell under the Bayesian sequential method is 1.988. Therefore, it is reasonable to compare the statistics of this method with $1-4$ fixed sample Bayesian hypothesis testing. Five metrics are considered: the final achieved maximum uncertainty $H_{\max,t_f}$; the final value for the cost function $\mathscr{J}(t_f)$; the total number of missed detections n_m; the total number of false detections n_f; and the total number of incorrect classifications n_i. For each case, 100 runs are carried out. For the sake of comparison, same settings are used for object number, positions, properties and initial position of the vehicle. All the other parameters are as in Section 5.3.5.1.

Figures 5.8(a)-5.8(e) show the performance comparison of the five metrics, respectively, between the fixed sample Bayesian hypothesis testings with 1,2,3,4 observations and the Bayesian sequential detection. Table 5.2 summarizes the statical results. In order to achieve similar small amount of decision errors, the fixed-sample hypothesis testing method requires $L=4$ observations at each cell. The risk-based sequential decision-making strategy outperforms the classical methods by 1) reducing decision errors, and 2) minimizing observation numbers. Therefore, according to Equations (5.13) and (5.14), under UCA, the proposed strategy leads to minimum Bayes risk within a same performance level.

Table 5.2 Performance comparison.

	$H_{\max,t_f}$	$\mathscr{J}(t_f)$	n_m	n_f	n_i
$L=1$	1.68E-2	2.85E-3	64.49	30.99	20.51
$L=2$	5.08E-3	1.7E-4	18.72	12.09	8.48
$L=3$	0	2.55E-5	12.02	7.43	2.91
$L=4$	0	4.39E-6	6.41	4.63	2.74
Sequential	1.18E-3	8.86E-5	9.07	3.98	2.15

5.4 Application to Space Situational Awareness

This section examines the problem of detecting and classifying objects in Earth orbit using a Space-Based Space Surveillance (SBSS) network. A SBSS system uses a combination of ground- and space-based sensors to monitor activities over a range of space orbits from low earth orbits up to an altitude higher than the geosynchronous orbit. The ternary risk-based sequential decision-making strategy developed in Section 5.3 is applied to object detection and classification using multiple range-angle sensors with intermittent information-sharing. The objective is to determine whether an object exists at a certain location (a cell in a discretization of the search space) or not, and, if an object exists, what type it belongs to. This is a nontrivial extension since, firstly, both the space-based sensors and the objects of interest are now constantly in orbital motion. Secondly, the search space is non-cartesian and will be discretized using a polar parametrization. Thirdly, the results for a single

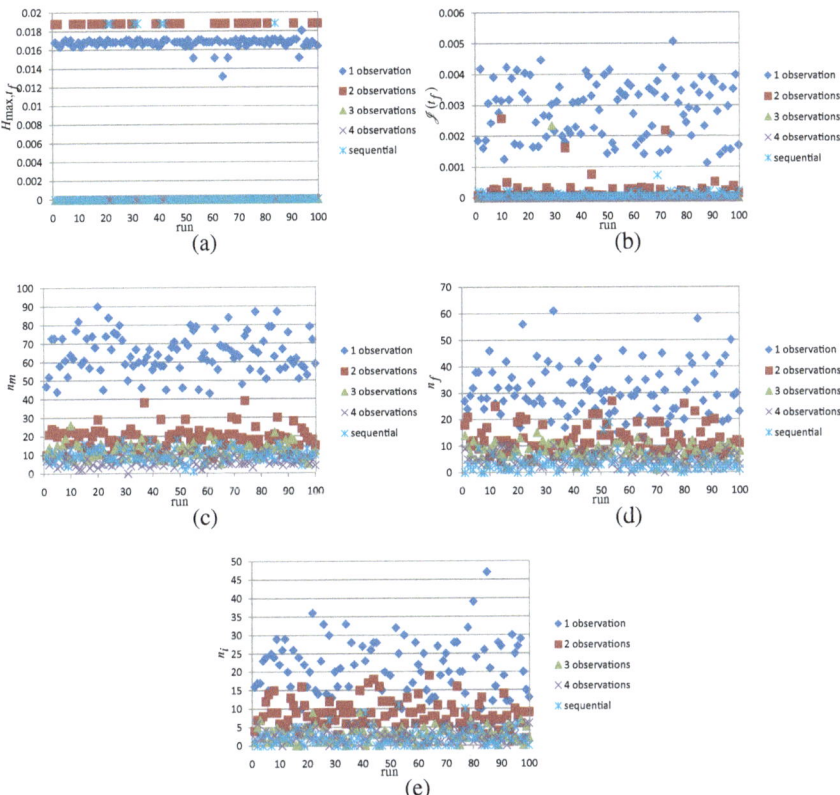

Fig. 5.8 Performance comparison.

sensor vehicle in Section 5.3 are extended to a SBSS network in which multiple sensors share information intermittently whenever sensors come within each other's communication range.

The problem is formulated in a simplified two-dimensional setting where the SBSS system is composed of four ground-based sensors and a space-based orbiting sensor satellite. This is done in order to reduce computational complexity while retaining the basic nontrivial elements of the problem. It will be shown that direct application of the proposed scheme will result in perfect detection and classification results for any object that exists in a geosynchronous orbit as long as it (at least) intermittently penetrates the field-of-regard of at least one sensor in the SBSS network. This is because, as observed in an earth-fixed coordinate frame, objects in geosynchronous orbit appear to be immobile. For objects in non-geosynchronous orbits, the assumption of immobility no longer holds and performance of the proposed approach significantly degrades.

5.4.1 Literature Review on SSA

Space Situational Awareness (SSA), that is, the monitoring of activities surrounding in- or through-space operations and the assessment of their implications, has received a great deal of attention in recent years, which was motivated initially by the publication of the Rumsfeld Commission Report [111]. More recently, the needs to keep track of all objects orbiting Earth has greatly increased due to the desire to prevent collisions, increased radio frequency interference, and limited space resources. NASA wants all objects as little as 1 cm to be tracked to protect the International Space Station, which would increase the number of tracked object from 10,000 to over 100,000 [3].

There are multiple decompositions of what SSA represents; from a capabilities point of view, SSA includes such things as:

* the ability to detect and track new and existing space objects to generate orbital characteristics and predict future motion as a function of time;
* monitoring and alert of associated launch and groundsite activities;
* identification and characterization of space objects to determine country of origin, mission, capabilities, and current status/intentions;
* understanding of the space environment, particularly as it will affect space systems and the services that they provide to users; and
* the generation, transmission, storage, retrieval, and discovery of data and information produced by sensor systems, including appropriate tools for fusion or correlation and the display of results in a form suitable for operators to make decisions in a timeframe compatible with the evolving situation

An excellent summary of the current system used by the United States to perform the detection and tracking functions of SSA, the Space Surveillance Network (SSN), is contained in [88], which includes current methods for tasking the network as well as proposed improvements.

5.4.2 System Model and Dynamics

System Model

Assume a uniform, spherical Earth. Figure 5.9 shows an example of the planar orbital sensor platform for the detection and classification of space objects used in this section.

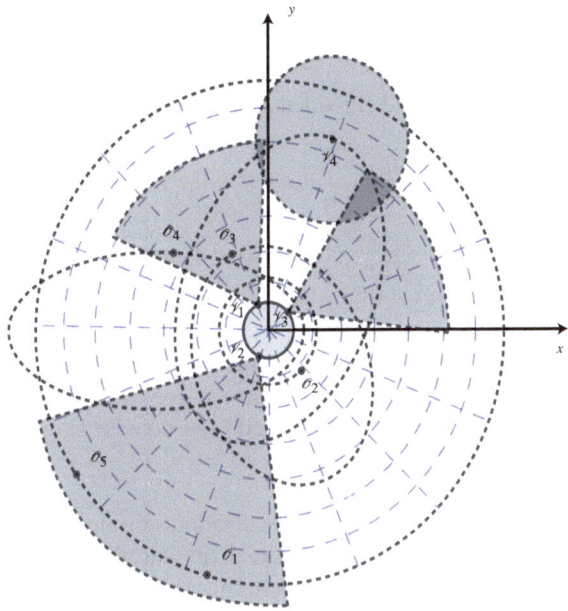

Fig. 5.9 Planar model of orbital sensor platform.

Consider a network of N_a sensors and N_o objects. Let $\mathscr{S} = \{\mathscr{V}_1, \mathscr{V}_2, ..., \mathscr{V}_{N_a}\}$ represent the set of sensors, that is, an entity that will accept the detection and classification tasks and will produce data and information. Let $\mathscr{O} = \{\mathscr{O}_1, \mathscr{O}_2, ..., \mathscr{O}_{N_o}\}$ represent the set of objects, that is, an entity that is not controllable or able to be tasked, and furthermore which it is desired to establish information about.

The ground-based sensors are stationary with respect to an earth-fixed frame. The dynamics of motion for ground-based sensors are as follows:

$$\dot{r}_i^s = 0, \tag{5.18}$$
$$\dot{\theta}_i^s = \omega_E,$$

where r_i^s and θ_i^s are the polar coordinates centered at the Earth for sensor i, and ω_E is the Earth's angular velocity. The space-based sensors follow Keplerian motion with the dynamics in polar form given by

$$\dot{r}_i^s = \sqrt{\frac{\mu}{a_i^s(1-(e_i^s)^2)}} e_i^s \sin(\theta_i^s - \omega_i^s), \tag{5.19}$$
$$\dot{\theta}_i^s = \sqrt{\frac{\mu}{a_i^s(1-(e_i^s)^2)}} \frac{1 + e_i^s \cos(\theta_i^s - \omega_i^s)}{r_i^s},$$

where μ is the Earth's gravitational parameter and equals to $398,600\text{km}^3/\text{s}^2$, a_i^s is the semi-major axis, e_i^s is the eccentricity, ω_i^s is the argument of perigee, and $\theta_i^s - \omega_i^s$ gives the true anomaly.

All objects to be detected and classified are assumed to be in orbit, and thus

$$\dot{r}_j^o = \sqrt{\frac{\mu}{a_j^o(1-(e_j^o)^2)}} e_j^o \sin(\theta_j^o - \omega_j^o), \tag{5.20}$$

$$\dot{\theta}_j^o = \sqrt{\frac{\mu}{a_j^o(1-(e_j^o)^2)}} \frac{1+e_j^o\cos(\theta_j^o - \omega_j^o)}{r_j^o}, \ j \in \mathcal{O}.$$

Here the mission domain $\mathscr{D} \subset \mathbb{R}^2$ is defined as the planar space domain from the Earth's surface up to an altitude higher than the geosynchronous orbit in which objects to be found and classified are located. The domain is discretized in polar coordinates as shown in Figure 5.9. Define $\mathbf{r}_j^o = (r_j^o, \theta_j^o)$ as the polar position of object j.

Here we focus on the detection and classification of objects located in geosynchronous orbits, and hence the ternary state $X(\check{\mathbf{c}})$ introduced in Section 5.3.1 is invariant with respect to time. For objects not in geosynchronous orbit, $X(\check{\mathbf{c}})$ will change with time as objects enter and leave cells. Hence, the actual state with respect to every cell $\check{\mathbf{c}}$ becomes a random process. To emphasize this time dependence, the state will be denoted by $X_t(\check{\mathbf{c}})$.

Sensor Model

Assume that the sensors are simple range-angle sensors [34]. First define

$$\rho(\mathbf{r}_i^s, \mathbf{r}_j^o) = \|\mathbf{r}_i^s - \mathbf{r}_j^o\|, \tag{5.21}$$

$$\psi(\mathbf{r}_i^s, \mathbf{r}_j^o) = \cos^{-1}\left(\frac{\mathbf{r}_i^s \cdot (\mathbf{r}_i^s - \mathbf{r}_j^o)}{r_i^s \|\mathbf{r}_i^s - \mathbf{r}_j^o\|}\right).$$

For the sake of brevity, the following shorthand notation will be used $\rho_{ij} \triangleq \rho(\mathbf{r}_i^s, \mathbf{r}_j^o)$ and $\psi_{ij} \triangleq \psi(\mathbf{r}_i^s, \mathbf{r}_j^o)$.

For each sensor $i \in \mathscr{A}$, define its maximum range as Υ_i and its maximum angle span as Ψ_i. The sensors are restricted to generate data only within a limited field-of-regard, e.g., an area around the sensor's position that it can effectively detect and classify objects within. Denote this area as Γ_i and define its boundary as the area swept out by a ray of length Υ_i relative to the sensor's current position and an angle Ψ_i measured in both directions from the local vertical direction at the sensor location. Thus

$$\Gamma_i = \{\mathbf{r} = (r, \theta) : \rho(\mathbf{r}_i^s, \mathbf{r}) \leq \Upsilon_i \text{ and } \psi(\mathbf{r}_i^s, \mathbf{r}) \leq \Psi_i\}. \tag{5.22}$$

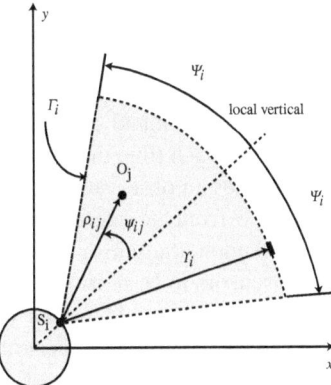

Fig. 5.10 Model for the range-angle sensor.

These quantities are illustrated in Figure 5.10. For ground-based sensors, which are limited by the local horizon, $-\frac{\pi}{2} \leq \Psi_i \leq \frac{\pi}{2}$. For space-based sensors, assuming they are allowed to arbitrarily re-orient their sensor payloads, would allow $-\pi \leq \Psi_i \leq \pi$. Each sensor is assumed to have a ternary discrete probability distribution within its sensory area Γ_i. Same detection probability is assumed everywhere within Γ_i. Despite this, the simple range-angle sensor model presented here is consistent with the limited-range vision-based sensor model considered in Section 2.2.3.1.

Communication Model

Two sensors can communicate with each other if they are within the communication region of one another and a line of sight between them exists. The neighbors of a sensor i are all sensors within the communication region Γ_i^c of i. Γ_i^c can be modeled in a similar way as the sensor's field-of-regard Γ_i given by Equation (5.22). Assume that the communication link is error free whenever a channel is established.

We make the following assumption. Whenever a communication link between two sensors is established, each sensor is assumed to have access to all the current observations from its neighboring sensors. Any previous observation from sensor i's neighbors in set $\mathcal{G}_i(t)$ at the current time step does not contribute to the state estimate associated with it at that time instant. The sensor updates its state estimate through data fusion (to be discussed soon) and makes a decision based on the posterior. Another fusion technique that one can apply is the decision fusion approach [128, 14]. Each sensor sends its neighbors a local decision derived by independent processing of its own observation. Some optimal decision rule is then used to fuse these local decisions. Due to the relatively lower amount of data to be transmitted, the decision fusion technique results in lower communication cost and higher data quality.

5.4.3 Decision-Making for Detection and Classification in Space

The ternary Bayesian sequential risk analysis developed in Section 5.3 is used as the decision-making strategy for detection and classification of space objects. The observation cost $c_{\text{obs}} > 0$ is assigned each time the sensor makes a new observation. This is because when a sensor makes an observation it is active and that withdraws power, which is a valuable resource, from the satellite. When all cells within a sensor domain are satisfactorily decided upon, the sensor can then be put in standby mode to save energy. If allow for the sensors to be non-omnidirectional and have control over the look direction of the sensor, c_{obs} should include both energy costs and costs associated with observing one group of cells at the cost of ignoring others.

Bayes' rule is employed to update the probability of object absence ($X(\tilde{\mathbf{c}}) = 0$), object having Property 'F' ($X(\tilde{\mathbf{c}}) = 1$), or object having Property 'G' ($X(\tilde{\mathbf{c}}) = 2$) associated with a particular sensor $\mathcal{V}_i$ at cell $\tilde{\mathbf{c}}$, based on observation taken by sensors in the set $\mathcal{G}_i(t)$ through intermittent communications.

Consider the Bayesian probability update equations given an observation sequence $\bar{Y}_t^i(\tilde{\mathbf{c}}) = \{\mathcal{V}_j \in \mathcal{G}_i(t) : Y_{j,t}(\tilde{\mathbf{c}}), \}$ available to sensor i at time step t. According to Bayes' rule, for each $\tilde{\mathbf{c}}$, it follows that

$$P_i(X(\tilde{\mathbf{c}}) = k | \bar{Y}_t^i(\tilde{\mathbf{c}}); t+1) = \eta_i P_i(\bar{Y}_t^i(\tilde{\mathbf{c}}) | X(\tilde{\mathbf{c}}) = k) P_i(X(\tilde{\mathbf{c}}) = k; t), \ k = 0, 1. (5.23)$$

Section 2.4.5.1 in Chapter 2 gives the detailed derivation and final expression for the above Bayes update equation. The information entropy function H_i (5.16) is used to measure the uncertainty level of object detection and classification. Here the subscript i is used to indicate that this level of uncertainty is associated with vehicle sensor $\mathcal{V}_i$.

The ground-based sensors take observations at certain fixed cells within their sensory area, while the space-based sensors follow the motion dynamics given by Equation (5.19) and travel through different cells with time. When a space-based sensor $\mathcal{V}_i$ leaves a cell, whether it made a decision or not, the uncertainty level H_i at this cell remains constant until the sensor comes back when possible. This is repeated until the uncertainty of the cell is within a small neighborhood of zero, i.e, when the detection and classification task is completed.

5.4.4 Simulation Results

Figure 5.11 shows the initial deployment of the space system architecture used in this simulation. The Earth is indicated by the green solid disc located at the origin of the polar coordinate system. The radius of the geosynchronous orbit $r_{\text{GEO}} = 42,157$ km is represented by the green circle. Discretize the space extending from the Earth's surface up to an altitude of $43,629$ km into 120 cells as shown in the figure. One space-based sensor and four ground-based sensors are indicated by the blue stars. The magenta ellipse shows the orbital trajectory of the orbiting sensor 1. For

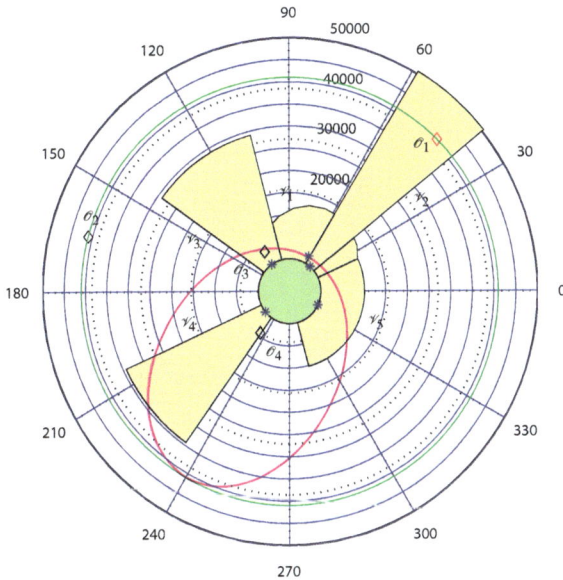

Fig. 5.11 Space system architecture.

the sake of simplicity in the simulation, it is assumed that $\Gamma_i^c = \Gamma_i$ for all sensors and are indicated by the yellow areas. The sensors communicate with each other and fuse their observations whenever they are within each other's communication region. The objects to be detected and classified are indicated by the diamond shapes, where the objects having Property 'F' are in black, and the object having Property 'G' is in red.

The orbital motions of the sensors and objects in the space system are simulated for 2 sidereal days. Figure 5.12(a) shows the probability of object 1 on geosynchronous orbit (cell 19) having property 'G' $P_1(X(\tilde{\mathbf{c}}) = 2|\bar{Y}_t^1; t+1)$ and its corresponding uncertainty function $H_1(P_1(\tilde{\mathbf{c}}_{19}, t))$ associated with the space-based sensor 1. Figure 5.12(b) shows $P_2(X(\tilde{\mathbf{c}}) = 2|\bar{Y}_t^2; t+1)$ and $H_2(P_2(\tilde{\mathbf{c}}_{19}, t))$ associated with the ground-based sensor 2. Because object 1 is constantly within the field-of-regard of sensor 2, the probability and uncertainty converge very quickly as shown by Figure 5.12(b). The space-based sensor 1 does not pass through cell 19 until after 1 day 11 hours and 37 minutes, hence the probability and uncertainty begin to evolve right after that time instant and also converge as shown by Figure 5.12(a).

Figure 5.13 shows the probability of object 2 on the geosynchronous orbit (cell 59) having property 'F' $P_1(X(\tilde{\mathbf{c}}) = 1|\bar{Y}_t^1; t+1)$ and its corresponding uncertainty function $H_1(P_1(\tilde{\mathbf{c}}_{59}, t))$ associated with the space-based sensor 1. Note that the space satellite 1 is the only sensor that can have view of $\mathcal{O}_2$ in the SBSS network. Because $\mathcal{O}_2$ enters its field-of-regard after 3 hours 50 minutes, the probability and uncertainty converge after that as shown by Figure 5.13. From the above results, it is shown that the objects on geosynchronous orbit can be detected and satisfactorily classified

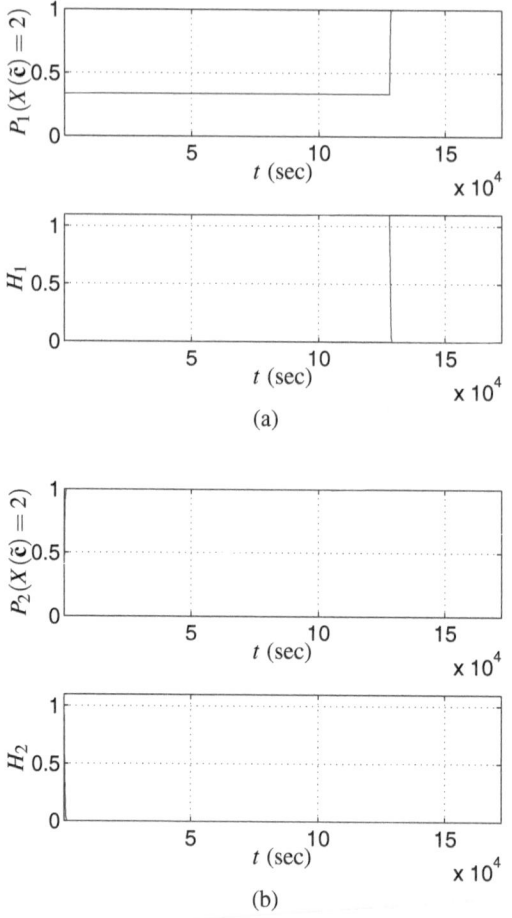

under the proposed approach because they appear to be immobile as viewed from an Earth-fixed frame.

Now investigate the performance for objects on non-geosynchronous orbits. For example, object 3 has entered and left cell 61 (within sensor 1's field-of-regard) and cell 41 (within sensor 3's field-of-regard) during the entire period. Figure 5.14(a) shows the probability of object absence $P_1(X(\tilde{c}) = 0|\bar{Y}_t^1; t+1)$ at cell 61 and the corresponding uncertainty function $H_1(P_1(\tilde{c}_{61}, t))$ associated with the space-based sensor 1. Figure 5.14(b) shows $P_3(X(\tilde{c}) = 0|\bar{Y}_t^3; t+1)$ and $H_3(P_3(\tilde{c}_{41}, t))$ associated with the ground-based sensor 3 at cell 41. Because object 3 is not on GEO orbit, its

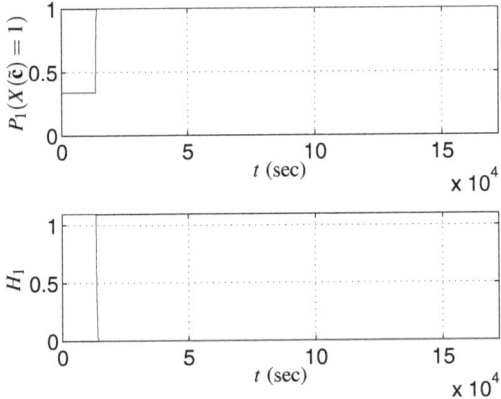

Fig. 5.13 Detection and classification results of $\mathcal{V}_1$ for $\mathcal{O}_2$.

position varies with respect to any discretized cell. The probability of object absence is decreased whenever an object passes through the cell within a sensor's field-of-regard and increases when the object is out of sight as shown by Figure 5.14. Once the probability of object absence approaches 1 at a cell, it will not decrease any more even if an object passes through it. Figure 5.15 shows similar results for object 4, which is also not on GEO orbit. Therefore, as anticipated, it is concluded that the proposed method does not guarantee good performance for the detection and classification of non-geosynchronous objects which are mobile as viewed from an Earth-fixed frame. To model the object mobility, a nonidentity transitional probability matrix for a dynamic Markov chain may be considered.

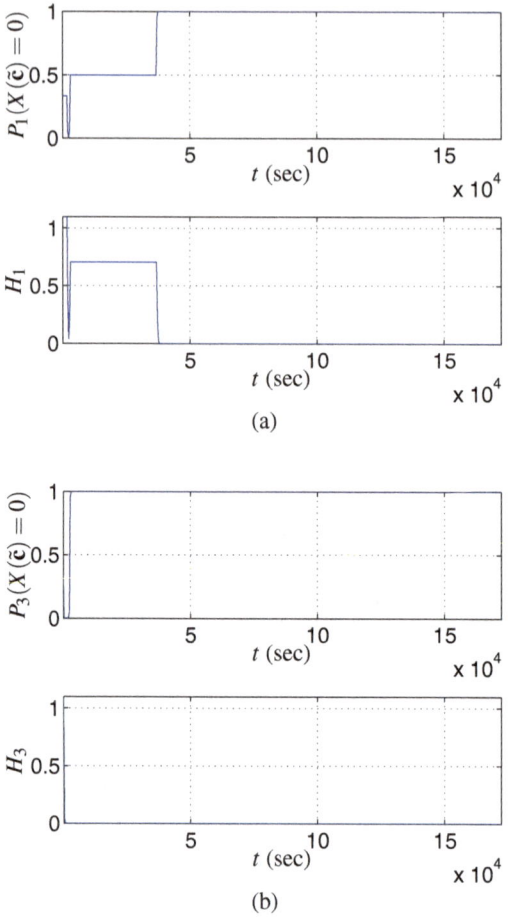

Fig. 5.14 Detection and classification results of $\mathscr{V}_1$ and $\mathscr{V}_3$ for $\mathscr{O}_3$.

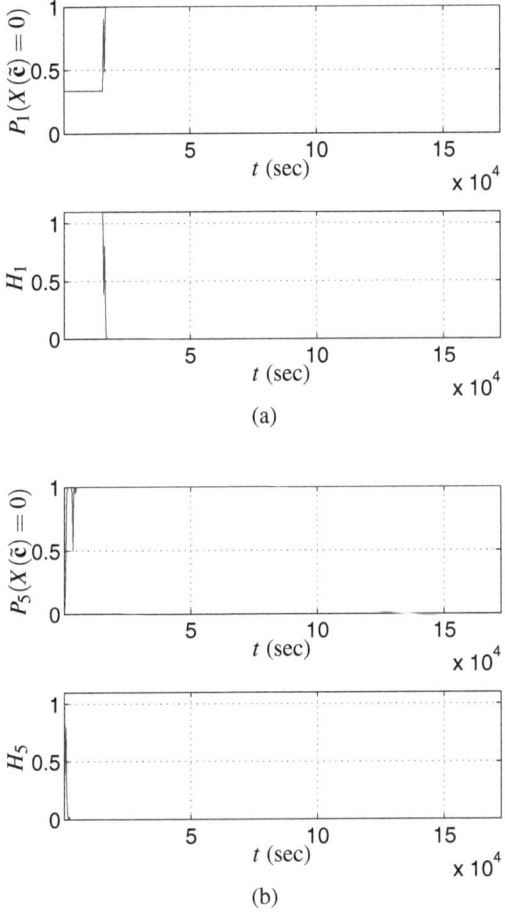

Fig. 5.15 Detection and classification results of $\mathcal{V}_1$ and $\mathcal{V}_5$ for $\mathcal{O}_4$.

Chapter 6
Risk-Based Sensor Management for Integrated Detection and Estimation

6.1 Introduction

In the realm of sensor network management, detection and estimation in the presence of uncertainties in both sensing and process dynamics are challenging tasks. Applications include but are not limited to using UAVs for fire detection and temperature estimation in aerial wild fire control [102], aerial search and tracking [82], space situational awareness (SSA) for the detection and categorizing of critical space objects [141], and chemical leak detection and concentration estimation in emergency responses to Chemical, Biological, Radiological and Nuclear, Explosive (CBRNE) incidents. In aerial wild fire control, for example, given limited sensing capabilities, the prompt detection of multiple distributed fire sources and the accurate estimation of the heat equation that governs the fire are key to mission success[1]. It is crucial to manage sensors in a way such that detection and estimation tasks are effectively assigned across search domain partitions and the detected processes to be estimated. This is especially true when the sensing resources are limited. More specifically, it is assumed that the sensors used for detection are the same ones used for estimation, albeit operated in different sensing modes. At every point in time, the sensor has to judge whether the currently available information is enough to make a detection or estimation decision. Once such a decision is made, a specification of where to search or what to estimate at the next time step has to be made. Considering the very limited sensory resources, these decisions need to be made in order to minimize risk, i.e., the optimal tradeoffs between the desired detection/estimation accuracy and the sensing costs paid to achieve it.

There is a rich literature on sensor management and task allocation. Among many other ad hoc architectures, one large category utilizes market-based auction

---

[1] If the process represents an object's position and velocity dynamics as in the domain search and object tracking problems, then it is assumed that the object can not leave its search domain partition. The results can be extended to the detection and tracking of mobile objects that move from one partition to another.

Y. Wang and I.I. Hussein: Search and Classification Using MAV, LNCIS 427, pp. 123–143.
springerlink.com   © Springer-Verlag London Limited 2012

algorithms for multi-robot coordination and task allocation (see [45, 118, 19] and references therein). In that literature, the proposed algorithms are deterministic rules assuming perfect sensing and communication links. The auction decisions across different tasks do not compete for sensory and/or communications resources. Another category of sensor management for multi-target tracking is driven by information theoretic measures [80, 70, 52]. The problem is formulated in a Bayesian framework and the sensor scheduling depends on the corresponding expected gain in information. However, the objective in these approaches is to maximize the expected information gain, or equivalently, to minimize the information uncertainty, by optimally selecting the targets to be tracked. Hence, the risk (i.e., the expected costs of the allocation decisions) associated with different sensing actions is not taken into account.

Chapter 5 investigates the problem of object search and classification treated as two competing tasks, which only requires the detection with respect to discrete random variables based on the assumption of stationary objects. For the integrated detection and estimation problem presented in this chapter, a single or multiple sensors are used to perform the detection of discrete random variables concurrently with the estimation of some other continuous random variables. First, the Bayesian sequential detection is utilized to address the detection problem. For estimation, the Bayesian sequential detection is extended to the Bayesian sequential estimation for continuous random variables [46]. The risk analysis for integrated detection and estimation requires the comparison of expected information gains for a hybrid mix of discrete (for detection) and continuous (for estimation) random variables. Here, the Rènyi information measures [108, 52] is used to model the information gained by making a certain sensor allocation decision. The relative information loss in making a suboptimal allocation decision is used to define the dynamic observation cost.

The main contribution of this chapter is *the integration of Bayesian sequential detection and estimation for a risk-based sensor management scheme given limited sensory resources and uncertainties in both state and observation models.*

6.2 Bayesian Sequential Detection

6.2.1 Problem Formulation

Denote the existence state as X, which is equal to 1 if a process exists within a given region and 0 if no process exists. The existence state X is modeled as a discrete-time, time independent Markov chain, where the transitional probability matrix is given by the identity matrix since it is assumed that the processes are with restricted mobility within the domain partition they occupy. Let Y_t be the observation random variable. The Bernoulli type sensor model in Section 2.4.1 is used with detection probability β.

Denote the probability of process existence by $\text{Prob}(X = 1; t) \equiv p_t$. Let $\bar{p}_{t+1} \equiv \text{Prob}(X = 1; t+1 | Y_{1:t})$ be the predicted conditional probability and $\hat{p}_t \equiv \text{Prob}(X = 1; t | Y_{1:t})$ be the updated conditional probability. The notation $\tilde{c}$ is omitted in $X(\tilde{c})$ with the understanding that the state is associated with an element (cell for detection in this section and object for estimation). Assuming identity transitional probability matrix, the following prediction step holds:

$$\bar{p}_{t+1} = \hat{p}_t. \tag{6.1}$$

At time t, the update step is as follows:

$$\hat{p}_t = \begin{cases} \frac{\beta \bar{p}_t}{(1-\beta)(1-\bar{p}_t)+\beta \bar{p}_t} & \text{if } Y_t = 1 \\ \frac{(1-\beta)\bar{p}_t}{\beta(1-\bar{p}_t)+(1-\beta)\bar{p}_t} & \text{if } Y_t = 0 \end{cases}. \tag{6.2}$$

This is consistent with Equation (2.33).

6.2.2 Bayesian Sequential Detection

The goal of Bayesian sequential detection is to determine the actual state of process existence X with minimum risk given a sequence of observations up to time t. The Bayesian sequential detection method in Chapter 5 is used here. Below, a brief review of the method is given, however, with the formulation consistent with the Bayesian sequential estimation method to be developed in Section 6.3. A set of simulation results are also provided to study the characteristics of the proposed method.

6.2.2.1 Decision Cost Assignment

First introduce the hypotheses: $\mathcal{H}_0$: the null hypothesis that $X = 0$; and $\mathcal{H}_1$: the alternative hypothesis that $X = 1$. Define the cost of accepting hypothesis $\mathcal{H}_i$ when the actual existence state is $X = j$ as C_{ij}. Using a Uniform Cost Assignment (UCA), the decision cost matrix is modified as follows

$$C_{ij} = \begin{cases} 0 & \text{if } i = j \\ c_d(\tau) & \text{if } i \neq j \end{cases}, \quad \tau \geq 0,$$

where $c_d(\tau) > 0$ is the cost of making the wrong detection decision at time $\tau \geq 0$ indicating the number of observations. To be consistent with the Bayesian sequential estimation method developed later, here the deterministic decision rule Δ is renamed as the detection estimator $\hat{X}_{t+\tau}$. It maps a sequence of observations $Y_{1:t+\tau}$ into a decision to accept $\mathcal{H}_0$ or $\mathcal{H}_1$, $\tau \geq 0$. Let the notation $C(\hat{X}_{t+\tau}(Y_{1:t+\tau}), X_{t+\tau})$ denote the cost of using estimator $\hat{X}_{t+\tau}$ given that the actual state of existence at time $t + \tau$ is $X_{t+\tau}$.

6.2.2.2   Detection Decision-Making

Restricting $\tau \leq 1$, there ends up to be six possible detection estimators and their corresponding Bayes risks r follow the procedures provided in Section 5.2.2:

$$r(\hat{X}_t^1, \tau = 0) = c_d(0)\hat{p}_t. \tag{6.3}$$

$$r(\hat{X}_t^2, \tau = 0) = c_d(0)(1 - \hat{p}_t). \tag{6.4}$$

$$r(\hat{X}_{t+1}^1(Y_{t+1}), \tau = 1) = c_d(1)\bar{p}_{t+1} + c_{\text{obs}}. \tag{6.5}$$

$$r(\hat{X}_{t+1}^2(Y_{t+1}), \tau = 1) = c_d(1)(1 - \bar{p}_{t+1}) + c_{\text{obs}}. \tag{6.6}$$

$$r(\hat{X}_{t+1}^3(Y_{t+1}), \tau = 1) = c_d(1)(1 - \beta) + c_{\text{obs}}. \tag{6.7}$$

$$r(\hat{X}_{t+1}^4(Y_{t+1}), \tau = 1) = c_d(1)\beta + c_{\text{obs}}. \tag{6.8}$$

The goal is to choose a combination of $\hat{X}_{t+\tau}$ and observation number τ that minimizes the Bayes risk. That is, the optimal decision is the one that gives the minimum risk:

$$r^*(\bar{p}_{t+\tau}) = \min_{\hat{X}_{t+\tau}, \tau} r(\hat{X}_{t+\tau}(Y_{t+\tau}), \tau).$$

Here, $r(\hat{X}_t^1, \tau = 0)$ and $r(\hat{X}_t^2, \tau = 0)$ correspond to making a detection decision at current cell without any further observation. Equations (6.5)-(6.8) correspond to postponing the decision and taking one more observation.

6.2.2.3   Simulation Results

In this section, the proposed optimal detection method is studied by varying the initial prior $\bar{p}_{t=0}$, sensor detection probability β, and observation cost c_{obs}, respectively. The actual state of existence is assumed to be $X = 1$ and random binary observations are taken. For every parameter choice, 30 simulations were run.

Varying Initial Prior Probability

Figure 6.1(a) shows the minimum Bayes risk curve with $\beta = 0.6$, $c_{\text{obs}} = 0.05$ and $c_d(0) = 1$, $c_d(1) = 0.3$ under different choices of initial priors $\bar{p}_0 = 0.2$, 0.5, 0.7. Figure 6.1(b) shows the updated probability $\hat{p}_t$ as a function of time with initial prior probability 0.2 (red), 0.5 (blue), and 0.7 (green), respectively. The two magenta horizontal lines correspond to the threshold probabilities π_L and π_U. Table 6.1 summarizes the statistical results of the simulation. Note that the minimum Bayes risk is the same in all the cases because the initial prior probability does not affect the value of the Bayes risk functions. Comparing the results, it can be seen that with a relatively better knowledge of X initially, the number of missed detections is lower and optimal decisions are made faster on average.

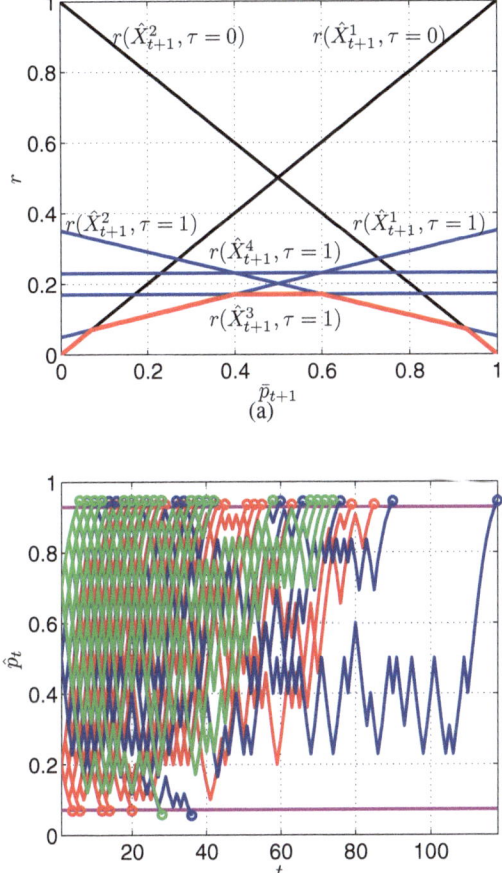

Fig. 6.1 Minimum Bayes risk curve and updated probability.

Table 6.1 Varying initial prior probability.

$\bar{p}_0$	π_L	π_U	$E[\hat{p}_t]$	Prob(missed detection)	Avg. observations
0.2	0.0714	0.9286	0.7330	16.67%	31.7667
0.5	0.0714	0.9286	0.9151	3.33%	33.0667
0.7	0.0714	0.9286	0.9169	3.33%	29.1333

Varying Sensor Detection Probability

Table 6.2 summarizes the statistical results of varying β. Other parameters are set as $\bar{p}_0 = 0.5$, $c_{obs} = 0.05$ and $c_d(0) = 1$, $c_d(1) = 0.3$. As can be seen in Table 6.2, a sensor with a very high detection probability ($\beta = 0.8$) or a very low detection probability ($\beta = 0.3$) outperforms a sensor with a value of β close to 0.5 and makes an optimal decision faster on average. This is because the proposed method depends on the sensor model. If the sensor quality is low, the optimal decision will be to accept the hypothesis that is opposite to the observed value of Y. However, if the detection probability is close to 0.5, i.e., the sensor returns a true or false observation with equal probability, more observations need to be taken before an optimal decision with minimum risk can be reached.

Table 6.2 Varying sensor detection probability.

β	π_L	π_U	$E[\hat{p}_t]$	Prob(missed detection)	Avg. observations
0.3	0.0714	0.9286	0.9362	3.33%	11.0667
0.6	0.0714	0.9286	0.8854	6.67%	35.8000
0.8	0.0714	0.9286	0.9118	3.33%	3.7333

Varying Observation Cost

Table 6.3 summarizes the statistical results of varying c_{obs}. Other parameters are set as $\bar{p}_0 = 0.5$, $\beta = 0.6$ and $c_d(0) = 1$, $c_d(1) = 0.3$. As can be seen in Table 6.3, with lower observation cost, the threshold probability is close to either 0 or 1, which implies that the sensor tends to take more observations until it reaches higher confidence level and ends up with more correct optimal decisions on average.

Table 6.3 Varying observation cost.

c_{obs}	π_L	π_U	$E[\hat{p}_t]$	Prob(missed detection)	Avg. observations
0.01	0.0143	0.9857	0.9886	0%	51.0000
0.05	0.0714	0.9286	0.8854	6.67%	27.4000
0.1	0.1429	0.8571	0.7558	16.67%	23.6667

6.3   Bayesian Sequential Estimation

6.3.1   *System Model: Single Sensor and a Single Process*

In this section Bayesian risk analysis tools is developed for sequential Bayesian estimation. Consider a linear system for a continuous random variable, which satisfies the discrete-time Markov chain model:

$$\mathbf{x}_{t+1} = \mathbf{F}_t \mathbf{x}_t + \mathbf{v}_t,$$
$$\mathbf{y}_t = \mathbf{H}_t \mathbf{x}_t + \mathbf{w}_t,$$

where $\{\mathbf{x}_t \in \mathbb{R}^n, t \in \mathbb{N}\}$ defines the process state sequence, $\mathbf{F}_t \in \mathbb{R}^{n \times n}$ is the process state matrix, $\{\mathbf{v}_t \in \mathbb{R}^n, t \in \mathbb{N}\}$ is the i.i.d. Gaussian process noise sequence with zero mean and positive semi-definite covariance $\mathbf{Q}_t \in \mathbb{R}^{n \times n}$, $\{\mathbf{y}_t \in \mathbb{R}^m, t \in \mathbb{N}\}$ is the measurement sequence, $\mathbf{H}_t \in \mathbb{R}^{m \times n}$ is the output matrix, and $\{\mathbf{w}_t \in \mathbb{R}^m, t \in \mathbb{N}\}$ is the i.i.d. Gaussian measurement noise sequence with zero mean and positive definite covariance $\mathbf{R}_t \in \mathbb{R}^{m \times m}$. The initial condition for the process state is assumed Gaussian with mean $\bar{\mathbf{x}}_0$ and positive definite covariance $\mathbf{P}_0 \in \mathbb{R}^{n \times n}$. It is assumed that the initial process state, process noise, and measurement noise are all uncorrelated.

6.3.2   *Sequential State Estimation*

In sequential estimation decision-making, it will be assumed that a suitable estimator has been constructed (here will use the Kalman filter) and the only decision to be made is whether to accept the estimate as the true state (and, hence, stop taking additional measurements) or to take (at least) one more measurement. Hence, the list of decisions are: (1) accept the estimate and stop taking measurements, and (2) take one more measurement.

6.3.3   *The State Estimation Problem*

For the estimation problem, the Kalman filter will be used since it is the optimal filter for linear Gaussian systems. At time step t, the process state and error covariance matrix prediction equations are given by [44]

$$\bar{\mathbf{x}}_t = \mathbf{F}_{t-1} \hat{\mathbf{x}}_{t-1},$$
$$\bar{\mathbf{P}}_t = \mathbf{Q}_{t-1} + \mathbf{F}_{t-1} \hat{\mathbf{P}}_{t-1} \mathbf{F}_{t-1}^{\mathrm{T}}, \qquad (6.9)$$

where $\hat{\mathbf{x}}_{t-1}$ is the process state estimate update at time t given measurements up to time $t-1$ and $\hat{\mathbf{P}}_{t-1}$ is the error covariance update up to time $t-1$. The posterior state estimate is given by:

$$\hat{\mathbf{x}}_t = \bar{\mathbf{x}}_t + \mathbf{K}_t \left(\mathbf{y}_t - \mathbf{H}_t \bar{\mathbf{x}}_t \right), \tag{6.10}$$

and the posterior error covariance matrix $\hat{\mathbf{P}}_t$ is given by:

$$\hat{\mathbf{P}}_t = \left(\mathbf{I} - \mathbf{K}_t \mathbf{H}_t \right) \bar{\mathbf{P}}_t. \tag{6.11}$$

In the above equations $\mathbf{I}$ is the identity matrix of dimension $n \times n$ and $\mathbf{K}_t$ is the Kalman gain:

$$\mathbf{K}_t = \bar{\mathbf{P}}_t \mathbf{H}_t^{\mathrm{T}} \left(\mathbf{H}_t \bar{\mathbf{P}}_t \mathbf{H}_t^{\mathrm{T}} + \mathbf{R}_t \right)^{-1}. \tag{6.12}$$

6.3.3.1 Estimation Error Cost Assignment

Let $\mathbf{x}_t^e(\mathbf{y}_t)$ be an estimator, i.e., computed estimate, of the actual process state $\mathbf{x}_t$ based on observation $\mathbf{y}_t$. Omit the dependence on $\mathbf{y}_t$ for notational brevity. Define the cost of accepting the estimate $\mathbf{x}_t^e$ given the actual process state $\mathbf{x}_t$ as $C(\mathbf{x}_t^e, \mathbf{x}_t)$. Set $C(\mathbf{x}_t^e, \mathbf{x}_t) = c_e(\tau) \|\mathbf{x}_t^e - \mathbf{x}_t\|^2$ (quadratic cost with $c_e(\tau) > 0$ being some τ-dependent cost value and $\tau \geq 0$ indicating the number of future observations), or the Uniform Cost Assignment:

$$C(\mathbf{x}_t^e, \mathbf{x}_t) = \begin{cases} 0 & \|\mathbf{x}_t^e - \mathbf{x}_t\| \leq \varepsilon \\ c_e(\tau) & \|\mathbf{x}_t^e - \mathbf{x}_t\| > \varepsilon \end{cases}, \tag{6.13}$$

where $\varepsilon > 0$ is some preset small interval. Here, for $\mathbf{x}_t^e$, the updated Kalman Filter estimate $\hat{\mathbf{x}}$ is used.

6.3.3.2 Estimation Decision-Making

At time t, after making a measurement $\mathbf{y}_t$, if it is decided not to take any more measurements, the Bayes risk is defined as the expected value (over all possible realizations of the process state, conditioned on all previous measurements) of the cost of choosing the estimate $\hat{\mathbf{x}}_t$:

$$r(\hat{\mathbf{x}}_t, \tau = 0) = E_{\mathbf{x}_t | \mathbf{y}_{1:t}} [C(\hat{\mathbf{x}}_t, \mathbf{x}_t)] = \int C(\hat{\mathbf{x}}_t, \mathbf{x}_t) p(\mathbf{x}_t | \mathbf{y}_{1:t}) d\mathbf{x}_t. \tag{6.14}$$

If assuming a quadratic cost assignment, it follows that

$$r(\hat{\mathbf{x}}_t, \tau = 0) = \int c_e(0)\|\hat{\mathbf{x}}_t - \mathbf{x}_t\|^2 p(\mathbf{x}_t|\mathbf{y}_{1:t})d\mathbf{x}_t = \int c_e(0)\sum_{i=1}^{n}(\hat{x}_t^i - x_t^i)^2 p(\mathbf{x}_t|\mathbf{y}_{1:t})d\mathbf{x}_t$$
$$= c_e(0)\text{Tr}\left[\hat{\mathbf{P}}_t\right],$$

where $c_e(0) > 0$ is the estimation cost when the sensor does not take an observation (i,e., $\tau = 0$), and $\hat{x}_t^i$ and x_t^i are the ith component of $\hat{\mathbf{x}}_t$ and $\mathbf{x}_t$, respectively.

The (expected) risk associated with taking more observations ($\tau \geq 1$) also needs to be computed. Since there are no measurements over time period $t+1 : t+\tau$ yet, define the conditional risk, $R_{\mathbf{x}_{t+1:t+\tau}}(\hat{\mathbf{x}}_{t+\tau}(\mathbf{y}_{t+1:t+\tau}), \tau)$ over all possible measurement realizations over $t+1 : t+\tau$ given the process state $\mathbf{x}_{t+\tau}$ at time $t+\tau$ as

$$R_{\mathbf{x}_{t+1:t+\tau}}(\hat{\mathbf{x}}_{t+\tau}(\mathbf{y}_{t+1:t+\tau}), \tau))$$
$$= E_{\mathbf{y}_{t+1:t+\tau}|\mathbf{x}_{t+1:t+\tau}}[C(\hat{\mathbf{x}}_{t+\tau}(\mathbf{y}_{t+1:t+\tau}), \mathbf{x}_{t+\tau})] + \kappa\tau c_{\text{obs}}$$
$$= \int C(\hat{\mathbf{x}}_{t+\tau}(\mathbf{y}_{t+1:t+\tau}), \mathbf{x}_{t+\tau})p(\mathbf{y}_{t+1:t+\tau}|\mathbf{x}_{t+1:t+\tau})d\mathbf{y}_{t+1:t+\tau} + \kappa\tau c_{\text{obs}},$$

where $\kappa > 0$ is some scaling parameter. The Bayes risk is defined as the weighted conditional risk $R_{\mathbf{x}_{t+1:t+\tau}}$, weighted by the predicted density function $p(\mathbf{x}_{t+1:t+\tau}|\mathbf{y}_{1:t})$ at time $t+1 : t+\tau$:

$$r(\hat{\mathbf{x}}_{t+\tau}(\mathbf{y}_{t+1:t+\tau}), \tau)$$
$$= E_{\mathbf{x}_{t+1:t+\tau}|\mathbf{y}_{1:t}}\left[R_{\mathbf{x}_{t+1:t+\tau}}(\hat{\mathbf{x}}_{t+\tau}(\mathbf{y}_{t+1:t+\tau}), \tau)\right]$$
$$= \int R_{\mathbf{x}_{t+1:t+\tau}}(\hat{\mathbf{x}}_{t+\tau}(\mathbf{y}_{t+1:t+\tau}), \tau)p(\mathbf{x}_{t+1:t+\tau}|\mathbf{y}_{1:t})d\mathbf{x}_{t+1:t+\tau}$$
$$= \int_{-\infty}^{\infty} p(\mathbf{x}_{t+1}|\mathbf{x}_{t+2:t+\tau}, \mathbf{y}_{1:t+\tau})d\mathbf{x}_{t+1}\cdots p(\mathbf{x}_{t+\tau-1}|\mathbf{x}_{t+\tau}, \mathbf{y}_{1:t+\tau})d\mathbf{x}_{t+\tau-1}$$
$$C(\hat{\mathbf{x}}_{t+\tau}(\mathbf{y}_{t+1:t+\tau}), \mathbf{x}_{t+\tau})p(\mathbf{x}_{t+\tau}|\mathbf{y}_{1:t+\tau})d\mathbf{x}_{t+\tau}p(\mathbf{y}_{t+1:t+\tau}|\mathbf{y}_{1:t})d\mathbf{y}_{t+1:t+\tau} + \kappa\tau c_{\text{obs}}. \quad (6.15)$$

If choosing a quadratic error cost assignment, the Bayes risk is given by

$$r(\hat{\mathbf{x}}_{t+\tau}(\mathbf{y}_{t+1:t+\tau}), \tau) = \int c_e(\tau)\|\hat{\mathbf{x}}_{t+\tau} - \mathbf{x}_{t+\tau}\|^2 p(\mathbf{x}_{t+\tau}|\mathbf{y}_{1:t+\tau})d\mathbf{x}_{t+\tau} + \kappa\tau c_{\text{obs}}$$
$$= c_e(\tau)\text{Tr}\left[\hat{\mathbf{P}}_{t+\tau}\right] + \kappa\tau c_{\text{obs}}. \quad (6.16)$$

Note that all the information required to compute $\hat{\mathbf{P}}_{t+\tau}$ is available at time t.

If choosing a UCA, then there is no closed-form expression for r unless the dimension of the process state is one, in which case the Bayes risk is given by

$$r(\hat{\mathbf{x}}_{t+\tau}(\mathbf{y}_{t+\tau}), \tau) = c_e(\tau)\left(1 - \text{Erf}\left(\frac{\varepsilon}{2\sqrt{2\hat{\mathbf{P}}_{t+\tau}}}\right)\right) + \kappa\tau c_{\text{obs}}, \quad \tau = 0, 1, \quad (6.17)$$

where

$$\text{Erf}(\cdot) = \frac{2}{\sqrt{\pi}} \int_0^{(\cdot)} e^{-t^2} dt$$

is the error function and ε is an error bound as indicated in Equation (6.13). For higher-dimension process state under UCA, the computation of r can be performed using Monte Carlo approximation techniques.

Since the optimal filter under linearity and normality assumptions is already determined by the Kalman filter, the only parameter to be optimized over is the observation number τ. The optimal decision corresponds to a particular observation number τ^* that yields minimum Bayes risk:

$$\tau^* = \text{argmin}_\tau r(\hat{\mathbf{x}}_{t+\tau}, \tau).$$

Remark

- Note that the Bayes risk is evaluated over all possible future realizations of the state $\mathbf{x}_{t+1:t+\tau}$ since the current prior is a sufficient statistic [46].
- Under the quadratic cost assignment, since the Kalman filter is used for estimation, an expression for the estimation risk for $\tau \geq 1$ is easily obtained (Equation (6.16)). However, there is no general formula for detection risk. This is because the optimal estimator for Bayesian sequential detection is unspecified and dynamically chosen in real-time from multiple candidates based on observation values, and is itself a function of the uncertainty in the detection process.
- To be consistent with Bayesian sequential detection, only $\tau = 0, 1$ in Equation (6.16) or Equation (6.17) will be used for estimation.   •

6.3.3.3   Simulation Results

In this section, the Bayesian sequential estimation method is applied on a time-invariant linear process and the performance is studied by varying the process noise covariance $\mathbf{Q}$, measurement noise covariance $\mathbf{R}$, and observation cost c_{obs}, respectively. UCA is assumed with error bound $\varepsilon = 0.1$.

Varying Process Noise Covariance

Figure 6.2 shows the estimation error between the actual process state $\mathbf{x}$ and the state estimate $\hat{\mathbf{x}}$ under different process noise covariances, where the blue lines correspond to $\mathbf{Q} = 0.001$, the red lines correspond to $\mathbf{Q} = 0.03$, and the black lines correspond to $\mathbf{Q} = 0.5$. The initial mean and covariance of the state is $\bar{\mathbf{x}}_0 = 5$ and $\mathbf{P}_0 = 1$, respectively. The state matrix is $F = 0.8$, the output matrix is $\mathbf{H} = 1$, and the measurement noise covariance is $\mathbf{R} = 0.1$. The observation cost is $c_{\text{obs}} = 0.02$. The process estimate gets updated when the sensor takes an observation. In the figure, squares are used to indicate the time steps when observations are taken and switch to circles when observations are no longer being taken. Beyond the switching point, which is indicated by a star, the state estimate is simply propagated without any state updates (since no new observations are made). As seen from the figure, when the process noise is larger, more observations need to be taken before an estimation could be accepted as the actual state with minimum risk.

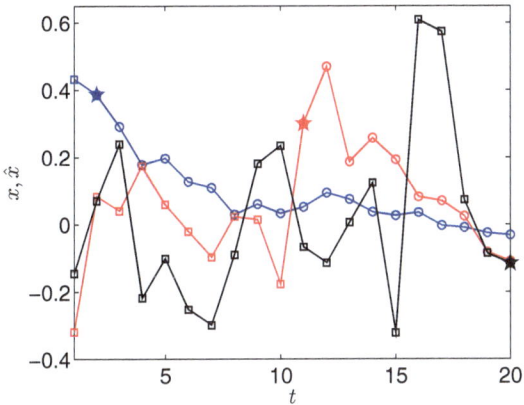

Fig. 6.2 Estimation error under different **Q**.

Varying Measurement Noise Covariance

Figure 6.3 shows the estimation error between the actual state **x** and the estimate of the process $\hat{\mathbf{x}}$ under different measurement noise covariances, where the blue line corresponds to $\mathbf{R} = 10$, the red line corresponds to $\mathbf{R} = 0.1$, and the black line

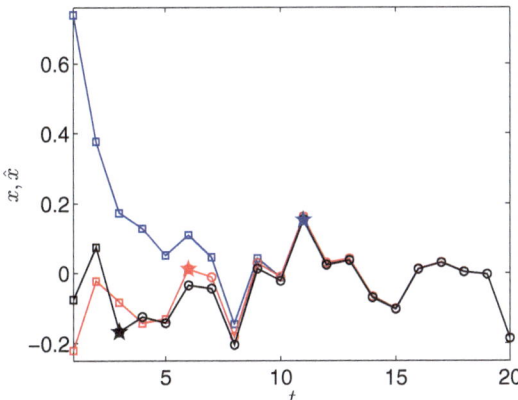

Fig. 6.3 Estimation error under different **R**.

corresponds to $\mathbf{R} = 0.01$. The initial mean and covariance of the state is $\bar{\mathbf{x}}_0 = 2$ and $\mathbf{P}_0 = 0.8$, respectively. The state matrix is $F = 0.7$, the output matrix is $\mathbf{H} = 1$, and the process noise covariance is $\mathbf{Q} = 0.01$. The observation cost is $c_{\text{obs}} = 0.01$. As seen from the figure, as the measurement noise gets larger, more observations need to be taken before an estimation with minimum risk is accepted as the optimal decision.

Varying Observation Cost

Figure 6.4 shows the estimation error between the actual state $\mathbf{x}$ and the estimate of the state $\hat{\mathbf{x}}$ under different observation costs, where the blue line corresponds to $c_{\text{obs}} = 0.001$, the red line corresponds to $c_{\text{obs}} = 0.01$, and the black line corresponds to $c_{\text{obs}} = 0.1$. The measurement noise covariance is $\mathbf{R} = 0.1$ and all other system parameters are the same as those in the case of varying $\mathbf{R}$. As seen from the figure, the sensor tends to take more observations before accepting an estimation as the true state when the observation cost is lower.

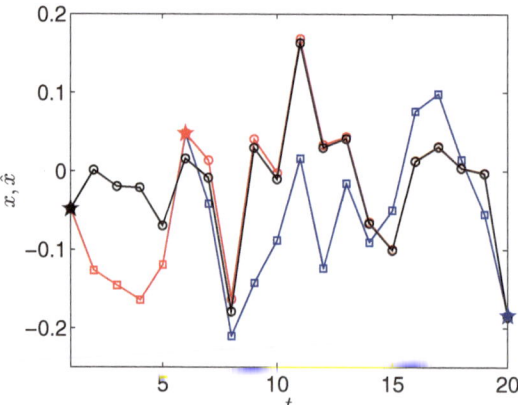

Fig. 6.4 Estimation error under different c_{obs}.

6.4   Extension to Multiple Elements

Now apply the Bayesian sequential detection for a discrete random variable in Section 6.2 to the detection of a possible process at cell $\tilde{\mathbf{c}}_j$ in the domain $\mathscr{D}$. That is, to determine if $X(\tilde{\mathbf{c}}_j) = 0$ or 1. Similarly, the Bayesian sequential estimation for a continuous random variable in Section 6.3 is applied to decide whether to accept the estimates (the updated process state $\hat{\mathbf{x}}_j$) of every detected process $\tilde{\mathbf{c}}_j$ if the existence state at cell $\tilde{\mathbf{c}}_j$ is $X(\tilde{\mathbf{c}}_j) = 1$. Here, the discrete state $X(\tilde{\mathbf{c}}_j) = 1$ could correspond, for example, to the existence of a fire in a forest domain cell with the continuous process to be estimated being a finite dimensional model of the diffusion equation within this cell.

First consider the Bayes detection risks at a cell $\tilde{\mathbf{c}}_j$. The risks associated with making a detection decision at $\tilde{\mathbf{c}}_j$ at the current time step t do not change in multi-element case because this is the decision associated with cell $\tilde{\mathbf{c}}_j$ itself. Hence, they are the same as Equations (6.3) and (6.4). Given that the sensor is observing $\tilde{\mathbf{c}}_j$ at t, the Bayes risk r_k associated with observing element $\tilde{\mathbf{c}}_k$ (including the possibility of choosing $\tilde{\mathbf{c}}_j$ again) at the next time step $t+1$ is defined as[2]:

$$r_k(\hat{X}_{k,t+1}(Y_{k,t+1}), \tau = 1) = E_{X_{k,t+1}|Y_{k,1:t}}[R_{X_{k,t+1}}(\hat{X}_{k,t+1}(Y_{k,t+1}), \tau = 1)], \quad (6.18)$$

where the conditional risk is given by:

$$R_{X_{k,t+1}}(\hat{X}_{k,t+1}(Y_{k,t+1}), \tau = 1) = E_{Y_{k,t+1}|X_{k,t+1}}[C(\hat{X}_{k,t+1}(Y_{k,t+1}), X_{k,t+1})] + c_{k,\text{obs}},$$

where $c_{k,\text{obs}}$ is the observation cost assigned for cell $\tilde{\mathbf{c}}_j$ if it decides to take an observation at element $\tilde{\mathbf{c}}_k$ at the next time step $t+1$. The optimal decision is then to choose a combination of $\hat{X}_{k,t+\tau}$, $\tau = 0, 1$, element $\tilde{\mathbf{c}}_k$ and observation number τ that minimizes Bayes risk:

$$r^*_{j,\min} = \min_{\hat{X}_{j,k,t+\tau},k,\tau} \left(r_j(\hat{X}_{j,t}, \tau = 0), r_k(\hat{X}_{k,t+1}(Y_{k,t+1}), \tau = 1) \right). \quad (6.19)$$

For the estimation of a detected process $\tilde{\mathbf{c}}_j$, the Bayes risk of not taking any more measurements is the same as Equation (6.14). Next, for process $\tilde{\mathbf{c}}_j$, compute the (expected) risk of taking one more measurement associated with some element $\tilde{\mathbf{c}}_k$:

$$r_k(\hat{\mathbf{x}}_{k,t+1}(\mathbf{y}_{k,t+1}), \tau = 1)$$
$$= \int \int C(\hat{\mathbf{x}}_{k,t+1}(\mathbf{y}_{k,t+1}), \mathbf{x}_{k,t+1}) p(\mathbf{x}_{k,t+1}|\mathbf{y}_{k,1:t+1}) d\mathbf{x}_{k,t+1} p(\mathbf{y}_{k,t+1}|\mathbf{y}_{k,1:t}) d\mathbf{y}_{k,t+1}$$
$$+ \kappa c_{k,\text{obs}}. \quad (6.20)$$

If under a quadratic cost assignment, the expected Bayes risk is given by

$$r_k(\hat{\mathbf{x}}_{k,t+1}(\mathbf{y}_{k,t+1}), \tau = 1) = c_e^k(1)\text{Tr}\left[\hat{\mathbf{P}}_{k,t+1}\right] + \kappa c_{k,\text{obs}},$$

where $c_e^k(1) > 0$ is the estimation cost with 1 observation associated with element k. If under UCA and assuming a 1 dimensional state, the Bayes risk is given by

$$r_k(\hat{\mathbf{x}}_{k,t+1}(\mathbf{y}_{k,t+1}), \tau = 1) = c_e^k(1)\left(1 - \text{Erf}\left(\frac{\varepsilon}{2\sqrt{2\hat{\mathbf{P}}_{k,t+1}}}\right)\right) + \kappa c_{k,\text{obs}}.$$

---

[2] Here the subscript j is added to emphasize the current cell $\tilde{\mathbf{c}}_j$ while the equations follow the same formulations as in Section 6.2.

The Bayesian sequential estimation method finds a particular combination of element $\tilde{c}_k$ and observation number ($\tau = 0$ or $\tau = 1$) that yields the decision with minimum Bayes risk $r^*_{j,\min}$ for each given observation.

$$r^*_{j,\min} = \min_{k,\tau} \left(r_j(\hat{x}_{j,t}, \tau = 0), r_k(\hat{x}_{k,t+1}, \tau = 1) \right). \tag{6.21}$$

6.5 Risk-Based Sensor Management

6.5.1 Problem Statement

In this section, a sensor management scheme is developed for integrated detection and estimation based on Bayesian sequential detection and estimation introduced in Sections 6.2 and 6.3 and their extension to multiple-element case in Section 6.4. Assume that a single sensor is capable of searching cells, and detecting and estimating processes, but not both at the same time. The Bayesian sequential detection and estimation methods are integrated into a unified risk analysis framework such that whenever a sensor chooses among multiple elements (cells for detection, processes for estimation), the resulting decision yields a minimum Bayes risk.

6.5.2 Detection and Estimation Sets

Let $Q_D(t) \subseteq \mathscr{D}$ be the set of cells for which no detection decision has been made up to time t (i.e., $r^*_{j,\min} \neq r_j(\hat{X}_{j,t}, \tau = 0)$ according to Equation (6.19)) and that are expected to be within the sensor's coverage area at the next time step $t+1$. Let $Q_T(t)$ be the set of detected processes ($X(\tilde{c}_j) = 1$) that still need further measurements for an acceptable estimate with minimum risk (i.e., $r^*_{j,\min} \neq r_j(\hat{x}_{j,t}, \tau = 0)$ according to Equation (6.21)) and that will be within the sensor's coverage area at the next time step $t+1$. Let $Q(t) = Q_D(t) \cup Q_T(t)$. Let $E(t)$ be the set of all cells in which it has been decided that no processes exist up to time t ($X(\tilde{c}_j) = 0$). Let $T(t)$ be the set of all processes that have the minimum Bayes risk based on all available observations up to time t and for which no further measurements are required (i.e., $r^*_{j,\min} = r_j(\hat{x}_{j,t}, \tau = 0)$ according to Equation (6.21)).

6.5.3 Decision List

At some time t, a sensor makes one of two types of measurements of an element $\tilde{c}_j$: (1) a detection measurement or (2) an estimation measurement. Based on the decisions made, an element $\tilde{\mathbf{c}}_j \in Q(t)$ (the grey dotted ellipse encompassing both Q_D and Q_T) can transition between the above mentioned sets at time t as shown in Figure 6.5.

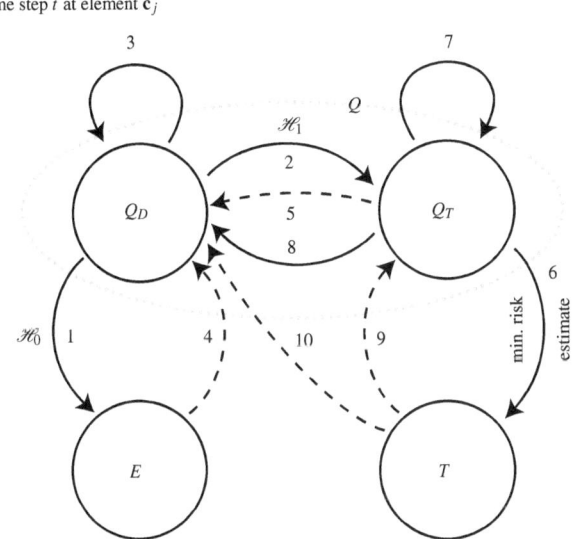

Fig. 6.5 Element transition.

In general, there are two main possible transitions:

- The current element $\tilde{\mathbf{c}}_j$ is a cell in $Q_D(t)$.

  1. Transition arrow 1: If no further observation is required and it is believed that the cell contains no process (hypothesis $\mathcal{H}_0$ is accepted), $\tilde{\mathbf{c}}_j$ is removed from $Q_D(t)$ and added to $E(t)$.
  2. Transition arrow 2: If no further observation is required and it is believed that the cell contains a process (hypothesis $\mathcal{H}_1$ is accepted), $\tilde{\mathbf{c}}_j$ is removed from $Q_D(t)$ and added to $Q_T(t)$ as a detected process that needs to be estimated.
  3. If more observations are required before making a detection decision, the sensor could either choose to take an observation a) at the current cell $\tilde{\mathbf{c}}_j$ (transition arrow 3), b) at another cell $\tilde{\mathbf{c}}_k \in Q_D(t+1)$ (transition arrow 3), or c) at another process $\tilde{\mathbf{c}}_k \in Q_T(t+1)$ (transition arrow 2). Note that the cell $\tilde{\mathbf{c}}_j$ still remains in $Q_D(t+1)$ at the next time step $t+1$.
  4. Also note that an element in E can transition back to Q_D (as indicated by the dashed transition arrow 4) if the previous detection result is no longer satisfactory. This also applies to an already detected process in Q_T (dashed transition arrow 5).

- The current element is a process $\tilde{\mathbf{c}}_j \in Q_T(t)$.

  1. Transition arrow 6: If no further observation is required, i.e., the process yields the minimum Bayes risk and the process estimate is accepted, $\tilde{\mathbf{c}}_j$ can be removed from $Q_T(t)$ and added to $T(t)$.
  2. If more observations are required before making an estimation decision, the sensor could either choose to take an observation a) at the current process $\tilde{\mathbf{c}}_j$ (transition arrow 7), b) at another process $\tilde{\mathbf{c}}_k \in Q_T(t+1)$ (transition arrow 7), or c) at another cell $\tilde{\mathbf{c}}_k \in Q_D(t+1)$ (transition arrow 8).
  3. If the estimation decision associated with process $\tilde{\mathbf{c}}_j$ does not give minimum Bayes risk any more, i.e., the process estimate can not be accepted as the true state any longer, this process is marked as "lost" and removed from $Q_T(t)$ and added to $Q_D(t)$ (dashed transition arrow 5). Moreover, as in detection, an element in $T(t)$ can transition back to $Q_T(t)$ (dashed transition arrow 9) or even Q_D (dashed transition arrow 10) directly if the previous estimated result is no longer acceptable.

At time step t, after taking an observation at an element $\tilde{\mathbf{c}}_j$, if $r_j^* = r_j^k(\tau = 1)$, $k \in Q(t)$, then it is less risky to take more observations than to stop detection or estimation at $\tilde{\mathbf{c}}_j$. The sensor is then allocated to element $\tilde{\mathbf{c}}_k$ at the next time step $t+1$.

6.5.4 Observation Decision Costs

The observation cost considered in this chapter is the relative loss of information gain that results from making a suboptimal sensor allocation decision. For each sensor allocation decision, associate with it a measure of gain in information. The decision yielding the maximum gain in information gives the optimal sensor allocation scheme and there is no loss. For each suboptimal decision, define the observation cost as the loss of gain in information relative to the optimum. Note that here suboptimal is in the sense of maximizing information gain only (e.g., not suboptimal with respect to risk minimization). Mathematically, the observation cost associated with element $\tilde{\mathbf{c}}_j$ is defined as

$$c_{j,\text{obs}} = E[I_{j^*}] - E[I_j], \tag{6.22}$$

where $E[I_j]$ is the expected information gain when measuring $\tilde{\mathbf{c}}_j$ and $\tilde{\mathbf{c}}_{j^*}$ is the element with the highest value of expected information gain.

The Rényi information divergence [108] will be used to compute the gain in information when comparing two probability densities, each belonging to either a cell (for detection) or a detected process (for estimation).

Rényi Information Divergence for Discrete Random Variables

For detection, the divergence is computed between two probability mass functions: the expected posterior probability mass function $\{\hat{p}_{j,t+1}, 1 - \hat{p}_{j,t+1}\}$ (given a measurement made at time $t+1$) and the predicted probability mass function $\{\bar{p}_{j,t}, 1 - \bar{p}_{j,t}\}$ [108]:

$$I_{j,\alpha}\left(\{\hat{p}_{j,t+1}, 1 - \hat{p}_{j,t+1}\} \mid \{\bar{p}_{j,t}, 1 - \bar{p}_{j,t}\}\right) = \frac{1}{\alpha - 1} \log_2 \left(\frac{\hat{p}_{j,t+1}^{\alpha}}{\bar{p}_{j,t+1}^{\alpha-1}} + \frac{(1 - \hat{p}_{j,t+1})^{\alpha}}{(1 - \bar{p}_{j,t+1})^{\alpha-1}} \right).$$

Here $\alpha = 0.5$ is used because this choice is reported as being most sensitive to the difference between two probability density functions [54].

If let $I_{j,\alpha;Y_{j,t+1}=1}$ and $I_{j,\alpha;Y_{j,t+1}=0}$ denote the Rényi information gain for the two possible types of sensor outputs at time $t+1$, the expected Rényi information gain is then given by

$$E_{Y_{j,t+1}|Y_{j,1:t}} I_{j,\alpha;Y_{j,t+1}} (\hat{p}_{j,t+1} \| \bar{p}_{j,t+1})$$

$$= \sum_{i=0}^{1} \text{Prob}(Y_{j,t+1} = i|Y_{j,1:t}) I_{j,\alpha;Y_{j,t+1}=i}$$

$$= \left[(1 - \beta)(1 - \bar{p}_{j,t+1|Y_{j,t+1}=1}) + \beta \bar{p}_{j,t+1|Y_{j,t+1}=1} \right] I_{j,\alpha;Y_{j,t+1}=1}$$

$$+ \left[\beta(1 - \bar{p}_{j,t+1|Y_{j,t+1}=0}) + (1 - \beta)\bar{p}_{j,t+1|Y_{j,t+1}=0} \right] I_{j,\alpha;Y_{j,t+1}=0}. \qquad (6.23)$$

Rényi Information Divergence for Continuous Random Variables

For estimation, the Rényi information divergence at time t is computed between two probability density functions: (a) the expected posterior probability density function $p(\mathbf{x}_{j,t+1}|\mathbf{y}_{j,1:t+1})$ after another (unknown) measurement $\mathbf{y}_{j,t+1}$ is made, and (b) the predicted density $p(\mathbf{x}_{j,t+1}|\mathbf{y}_{j,1:t})$ given the measurements up to $\mathbf{y}_{j,t}$ [108, 52]

$$I_{j,\alpha}\left(p(\mathbf{x}_{j,t+1}|\mathbf{y}_{j,1:t+1}) | p(\mathbf{x}_{j,t+1}|\mathbf{y}_{j,1:t})\right) \qquad (6.24)$$

$$= \frac{1}{\alpha - 1} \log_2 \int p(\mathbf{x}_{j,t+1}|\mathbf{y}_{j,1:t}) \left(\frac{p(\mathbf{x}_{j,t+1}|\mathbf{y}_{j,1:t+1})}{p(\mathbf{x}_{j,t+1}|\mathbf{y}_{j,1:t})} \right)^{\alpha} d\mathbf{x}_{j,t+1}.$$

For linear Guassian models combined with a Kalman filter, it follows that [52]:

$$E_{y_{j,t+1}|y_{j,1:t}}I_{j,\alpha}(p_j(\mathbf{x}_{t+1}|\mathbf{y}_{1:t+1})\|p_j(\mathbf{x}_{t+1}|\mathbf{y}_{1:t}))$$

$$= \frac{1}{2(1-\alpha)}\log\left(\frac{|\alpha\mathbf{R}_j^{-1}\mathbf{H}_j\bar{\mathbf{P}}_{j,t+1}\mathbf{H}_j^{\mathsf{T}}+\mathbf{I}|}{|\mathbf{R}_j^{-1}\mathbf{H}_j\bar{\mathbf{P}}_{j,t+1}\mathbf{H}_j^{\mathsf{T}}+\mathbf{I}|^{\alpha}}\right)+\frac{1}{2}\mathrm{Tr}\left[\mathbf{I}-\left(\alpha\mathbf{R}_j^{-1}\mathbf{H}_j\bar{\mathbf{P}}_{j,t+1}\mathbf{H}_j^{\mathsf{T}}+\mathbf{I}\right)^{-1}\right]$$

6.5.5 Solution Approach

Figure 6.6 summarizes the solution algorithm as a general flow chart. At time step t, the sensor takes an observation ($Y_{j,t}$ for detection or $\mathbf{y}_{j,t}$ for estimation) at the current element $\tilde{\mathbf{c}}_j \in Q(t-1)$. Based on this real-time observation and the prior probability/estimate ($\bar{p}_{j,t}$ for detection or $\bar{\mathbf{x}}_{j,t}$ for estimation), the updated (posterior) probability/estimate ($\hat{p}_{j,t}$ for detection and $\hat{\mathbf{x}}_{j,t}$ for estimation) and the predicted probability/estimate ($\bar{p}_{j,t+1}$ for detection and $\bar{\mathbf{x}}_{j,t+1}$ for estimation) are obtained via a recursive implementation (Bayesian update Equations (6.2) and (6.1) for detection and Kalman filter Equations (6.10)-(6.12) and (6.9) for estimation). Note that the predicted probability/estimate is treated as the prior probability/estimate at the next time step $t+1$. Then the corresponding Bayes risk are computed, where the updated probability/estimate is used to compute the Bayes risk $r_j(\tau = 0)$ of making a direct detection or estimation decision without taking any further observations (i.e., future observation length $\tau = 0$) (Equation (6.3) or (6.4) for detection and Equation (6.14) for estimation), and the predicted probability/estimate is used to compute the Bayes risk $r_j(\tau = 1)$ associated with taking one more observation ($\tau = 1$) for a possibly better decision (Equations (6.5-6.8) for detection and Equation (6.15) for estimation). Bayesian sequential decision-making is then employed as follows. If the minimum Bayes risk $r^*_{j,\min}$ is giving by taking future observations ($\tau = 1$), then the sensor will take an observation at some element $\tilde{\mathbf{c}}_k \in Q(t)$ (including the possibility of choosing $\tilde{\mathbf{c}}_j$) that minimizes the Bayes risk at the next time step $t+1$ (according to Equation (6.18) for detection and Equation (6.20) for estimation). Otherwise ($\tau = 0$), the sensor makes a detection or estimation decision at $\tilde{\mathbf{c}}_j$, and moves to some $\tilde{\mathbf{c}}_k \in Q(t)\backslash\{\tilde{\mathbf{c}}_j\}$ that minimizes the Bayes risk and takes an observation at that element at the next time step $t+1$ (Equation (6.18) for detection and Equation (6.20) for estimation). This process is repeated until a detection or estimation decision can be made at every element in $Q(t)$.

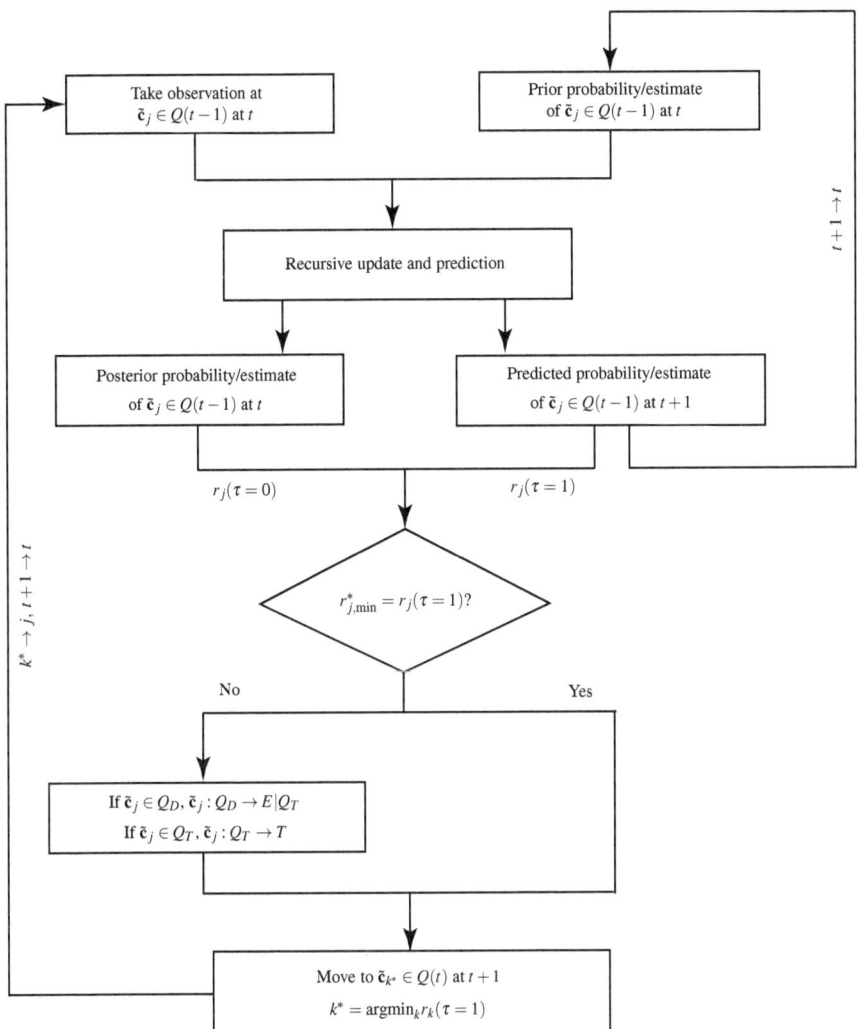

Fig. 6.6 Decision flowchart.

6.6 Simulation Results

Assume there are $N_{tot} = 10$ cells initially, among which there are 7 processes (Cell 1-7) to be detected and estimated. Both the number of processes and their cell numbers are unknown to the algorithm beforehand. A limited-range sensor is used, which is capable of taking either a detection or an estimation observation on any one cell or process at every time step. The initial predicted probability $\bar{p}_{j,t=0}$ for $j = 3$ is set to be 0.1 and that for all the other cells is 0.5. The value of the sensor detection probability β associated with each cell follows a Gaussian distribution with mean 0.6 and variance 0.1. The process states are assumed to be time-invariant Gaussian processes with zero mean and positive definite covariance 0.1. Same parameters are used for the processes: $\mathbf{F} = 1, \mathbf{H} = 1, \mathbf{R} = 1, \mathbf{Q} = 0.1$. For both detection and estimation, UCA is assumed and ε is set to be 0.1. The probability of the existence state or the estimate of the process state will be updated when the sensor decides to take a measurement of this element. When there is no observation, either because that a detection/estimation decision has been made at the current element or the sensor decides to postpone the decision and makes a measurement elsewhere, the probability/estimate is propagated based on the predictions. The decision costs for detection and estimation are $c_d(0) = 1, c_d(1) = 0.3, c_e(0) = 1, c_e(1) = 0.16$. The information gain scaling parameter κ is chosen to be 0.06.

Remark about Parameter Sensitivity. Simulation results were very sensitive to parameter choices. Some parameter choices lead to excessive detection observations and others to exhaustive estimations for a single detected process.   •

The results of running the algorithm until the stopping criteria is met, i.e., the detection decisions for all cells and estimation decisions for all detected processes are made with minimum Bayes risk, are shown in Figures 6.7 and 6.8. All the processes have been detected and satisfactorily estimated except that there is a missed detection at Cell 4. Figure 6.7 shows the assigned observing cell at each time step according to the proposed integrated decision-making strategy. The green dots represent the detection stopping time when the hypothesis $\mathcal{H}_1$ is accepted. The green squares indicate the detection stopping time when the alternate hypothesis $\mathcal{H}_0$ is accepted. For example, there is a missed detection at Cell 4, no estimation is performed after the detection decision $X_4 = 0$ is made at time step 137. Note that an already detected process can be estimated before other processes have been detected, however, a process must first be detected before being estimated. Figures 6.8(a), 6.8(b) and 6.8(c) show the actual probability $p_{j,t}$ (blue) and the updated probability $\hat{p}_{j,t}$ (red), the actual process state $\mathbf{x}$ (blue) and the estimate of the state $\hat{\mathbf{x}}$ (red) for Cell 4, 6, and 8, respectively. Figure 6.8(d) enlarges the estimation performance of Cell 1 during time period 1-300. The horizontal lines in the figure correspond to the moments when the sensor decides to stop taking observations and the estimates of the state propagate based on the predictions ($\mathbf{F} = 1$).

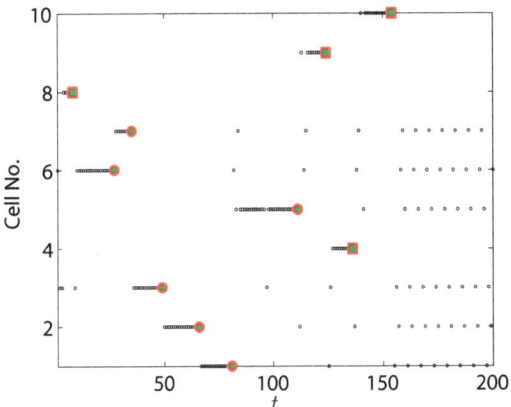

Fig. 6.7 Observed cell at each time step for the 10-cell problem.

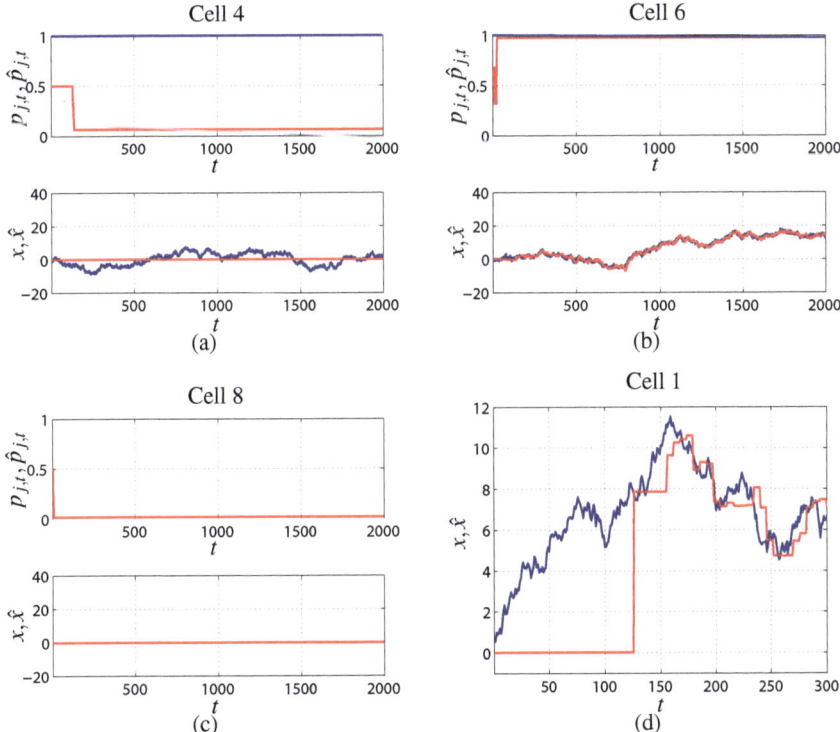

Fig. 6.8 Actual and updated probability, actual and estimate of the process state.

Chapter 7
Conclusion and Future Work

In this book, real-time decision-making strategies are investigated for domain search and object classification using MAVs under limited sensory resources. Because search requires for vehicle mobility and classification requires the vehicle to be in the vicinity of each found object of interest, these two tasks compete for the limited sensory resources, especially over large-scale mission domains. We formulate the domain search problem as a coverage control problem which aims at constructing a high-confidence awareness map of the entire mission domain. The detection of every object within the domain guarantees the completion of the search task. Based on this concept, Lyapunov-based, awareness-based, and Bayesian-based dynamic coverage control strategies are introduced in sequence. All the proposed coverage control strategies guarantee the detection of all objects inside the search domain. The proposed Lyapunov-based coverage control law is applied to seafloor mapping using multiple AUVs. The awareness-based coverage control law models the awareness level of MAVs of events occurring at very point in the domain. For the Bayesian-based coverage control strategy, we discretize the search domain and consider the uncertainties in sensor perception in a probabilistic framework.

Correspondingly, a deterministic awareness-based decision-making strategy for search and classification is developed, which guarantees the detection of all unknown objects of interest and the classification of each found object by at least a desired amount of time. In order to take into account sensor errors, a probabilistic Bayesian-based decision-making strategy is then developed. The states of object existence and its classification are treated as discrete random variables. Bayes filter is utilized and all the objects are shown to be detected and classified with 100% certainty. To further consider the cost of taking each new observation for a probably better decision, a risk-based decision-making strategy based on binary Bayesian sequential detection method is presented. The proposed strategy is guaranteed to complete both competing tasks with a minimum risk in a probabilistic framework. This binary decision-making strategy is further extended to more general ternary settings. The results are applied to the SSA problem in SBSS systems. Combining both Bayesian sequential detection and its extension to Bayesian sequential estimation,

Y. Wang and I.I. Hussein: Search and Classification Using MAV, LNCIS 427, pp. 145–148.
springerlink.com © Springer-Verlag London Limited 2012

an optimal risk-based sensor management scheme is proposed for integrated detection and estimation. It involves both hypothesis testing for discrete random variables and estimation for continuous random variables. Based on Bayesian sequential detection for discrete random variables, the results are extended to Bayesian sequential estimation for continuous random variables. Both parts are integrated into a unified risk-based decision-making scheme, which facilitates optimal resource allocation across multiple tasks that are competing for the same limited sensory resources. Simulation results are provided to illustrate the performance of the proposed strategies.

A summary of future research directions is as follows.

Search vs. Tracking Decision-Making for Mobile Objects

The current work considers the search and classification decision-making of stationary objects, which may also be extended to mobile objects whose motion trajectories are restricted in a discretized cell. However, more generalized mobile objects is of interest due to the wide applications on search and tracking problems. The mobility of the objects may be modeled according to Markov chains with non-identity transition probability matrix under the probabilistic framework. This technique can be used to develop strategies for the search and tracking of space objects on non-geosynchronous orbits in the SSA problem.

Sequential Risk Analysis

In this book, the domain search and object classification problems are considered in the Bayesian-based probabilistic framework. That is to say, we assume a prior known information about object distribution within the domain and the properties of a found object of interest. In cases where no prior information is available, SPRT, Neyman-Pearson, SR and CUSUM based hypothesis testings will be adopted for the risk analysis associated with decision-making. Both centralized and decentralized versions of Bayesian sequential detection and SPRT methods can be developed for sequential detection as well as estimation. The integration of these approaches will provide a general scheme for unified detection and estimation.

Applications on SSA

The proposed risk-based sensor management scheme may be applied to the SSA problem by incorporating nonlinear Keplerian spacecraft dynamics. As mentioned above, Markov chains with non-identity transition probability matrix may be used to model object mobility. However, computationally efficient (approximate) algorithms will be a necessity to tackle the issue of large amount of data raised in this case.

Vehicle Dynamics

Vehicle dynamics can also be taken into account into the system model. To be more specific, vehicle motion control strategies for second-order nonlinear vehicle dynamics including motion uncertainties and nonholonomic constraints can be considered. Application of the coverage control laws to underwater sea floor mapping can be modified to incorporate both vehicle dynamics and ocean fluid dynamics. Furthermore, the modifications of decision-making strategies considering higher-order dynamics and vehicle physical constraints will not be trivial.

MAV Decision Fusion

For MAV cooperative decision-making, besides the sensor fusion algorithm introduced in Section 2.4.5, the decision fusion technique offers a more affordable approach for MAV communications. This is because decision fusion only requires the transmission of a made decision from each cooperative vehicle instead of the relatively large amount of observation data for sensor fusion.

Nonlinear Systems

In aerospace applications, the governing equations are almost always nonlinear, as in the tracking of objects orbiting Earth. Therefore, estimating nonlinear orbital dynamics given limited sensing resources is an essential extension of the work presented in this book. The Bayesian sequential estimation method can be extended to nonlinear systems via, for example, Gaussian sum filters or particle filters.

Domain Discretization

In this book, it is assumed that the domain discretization is fine enough such that there is at most one object at a single cell. This assumption can be relaxed by allowing more than a single object per cell via target discrimination and data association.

Unknown Environment Geometries

Current work assumes mission domains with known geometries. The problem of unknown environment exploration is of interest for realistic implementations. This problem may be solved by predicting a vehicle sensor's position at the next time step, which will be utilized to estimate the dynamic search space.

Uncertainty in Vehicle Actions

Besides the uncertainty in sensor perception, the Partially Observable Markov Decision Process (POMDP) may be used to model the uncertainty in the outcomes of vehicle actions. POMDP method associates a cost function with each vehicle action and keeps updating the value function using dynamic programming. The solution of POMDP yields an optimal policy that maximizes the total expected rewards. For cases when a vehicle's action affects decisions made by others, game theory may be utilized to find dominant strategy for each individual vehicle or the equilibrium for MAVs.

References

1. Robomower, http://www.friendlyrobotics.com
2. Acar, E.U., Choset, H.: Sensor-Based Coverage of Unknown Environments: Incremental Construction of Morse Decompositions. International Journal of Robotics Research 21(4), 345–366 (2002)
3. Ailor, W.: Space Traffic Control: A View of the Future. Space Policy 18(2), 99–105 (2002)
4. Akella, S., Hutchinson, S.: Coordinating the Motions of Multiple Robots with Specified Trajectories. In: IEEE International Conference on Robotics and Automation, May 2002, pp. 624–631 (2002)
5. Alighanbari, M., How, J.: An Unbiased Kalman Consensus Algorithm. In: Proceedings of the American Control Conference, pp. 3519–3524 (2006)
6. Baillieul, J., Antsaklis, P.J.: Control and Communication Challenges in Networked Real-Time Systems. Proceedings of the IEEE 95(1), 9–28 (2007)
7. Basseville, M.E., Nikiforov, I.V.: Detection of Abrupt Changes: Theory and Application. In: Information and System Sciences. Prentice-Hall (April 1993)
8. Beard, R.W., McLain, T.W.: Multiple UAV Cooperative Search under Collision Avoidance and Limited Range Communicaiton Constraints. In: IEEE Conference on Decision and Control, December 2003, pp. 25–30 (2003)
9. Bellingham, J.S., Tillerson, M., Alighanbari, M., How, J.P.: Cooperative Path Planning for Multiple UAVs in Dynamic and Uncertain Environments. In: Proceedings of the IEEE Conference on Decision and Control, December 2002, pp. 2816–2822 (2002)
10. Benkoski, S., Monticino, M., Weisinger, J.: A Survey of the Search Theory Literature. Naval Research Logistics 38(4), 469–494 (1991)
11. Bertozzi, M., Broggi, A., Fascioli, A.: Vision-Based Intelligent Vehicles: State of the Art and Perspectives. Robotics and Autonomous Systems 32, 1–16 (2000)
12. Bertuccelli, L.F., How, J.P.: Robust UAV Search for Environments with Imprecise Probability Maps. In: Proceedings of the 44th IEEE Conference on Decision and Control, and the European Control Conference, December 2005, pp. 5680–5685 (2005)
13. Bertuccelli, L.F., How, J.P.: Bayesian Forecasting in Multi-vehicle Search Operations. In: AIAA Guidance, Navigation, and Control Conference and Exhibit (August 2006)
14. Blum, R.S., Kassam, A., Poor, H.V.: Distributed Detection with Multiple Sensors: Part 2 - Advanced Topics. Proc. IEEE 85(1), 64–79 (1997)
15. Cassandras, C.G., Li, W.: Sensor Networks and Cooperative Control. European Journal of Control 11(4-5), 436–463 (2005)

16. Chandler, P., Pachter, M., Nygard, K., Swaroop, D.: Cooperative Control for Target Classification. In: Murphey, R., Pardalos, P.M. (eds.) Cooperative Control and Optimization, pp. 1–19. Kluwer Academic Publishers (2002)
17. Chandler, P.R., Pachter, M., Rasmussen, S.: UAV Cooperative Control. In: Proceedings of the American Control Conference, June 2001, pp. 50–55 (2001)
18. Chen, S.Y., Li, Y.F.: Automatic Sensor Placement for Model-Based Robot Vision. IEEE Transactions on Systems, Man, and Cybernetics I- Part B: Cybernetics 34(1), 393–408 (2004)
19. Choi, H.-L., Brunet, L., How, J.P.: Consensus-Based Decentralized Auctions for Robust Task Allocation. IEEE Transactions on Robotics 25(4), 912–926 (2009)
20. Choset, H.: Coverage for Robotics: A Survey of Recent Results. Annals of Mathematics and Artificial Intelligence 31(1), 113–126 (2001)
21. Choset, H., Nagatani, K.: Topological Simultaneous Localization and Mapping (SLAM): Toward Exact Localization without Explicit Localization. IEEE Transactions on Robotics and Automation 17, 125–137 (2001)
22. Coifmana, B., Beymer, D., McLauchlan, P., Malik, J.: A Real-Time Computer Vision System for Vehicle Tracking and Traffic Surveillance. Transportation Research Part C: Emerging Technologies 6(4), 271–288 (1998)
23. Cortés, J., Martínez, S., Bullo, F.: Spatially Distributed Coverage Optimization and Control with Limited-Range Interactions. In: ESAIM. Control, Optimisation and Calculus of Variations, pp. 691–719 (2005)
24. Cortés, J., Martínez, S., Karatus, T., Bullo, F.: Coverage Control for Mobile Sensing Networks. IEEE Transactions on Robotics and Automation 20(2), 243–255 (2004)
25. Cover, T.M., Thomas, J.A.: Elements of Information Theory, 2nd edn. Wiley (2006)
26. Davids, A.: Urban Search and Rescue Robots: from Tragedy to Technology. IEEE Intelligent Systems 17(2), 81–83 (2002)
27. DeBolt, C., O'Donnell, C., Freed, C., Nguyen, T.: The BUGS (Basic UXO Gathering Systems) Project for UXO Clearance & Mine Countermeasures. In: IEEE International Conference on Robotics and Automation, April 1997, pp. 329–332 (1997)
28. Dissanayake, G., Newman, P., Clark, S., Durrant-Whyte, H.F., Csorba, M.: A Solution to the Simultaneous Localization and Map Building (SLAM) Problem. IEEE Transactions on Robotics and Automation 17(3), 229–241 (2001)
29. Drezner, Z.: Facility Location: A Survey of Applications and Methods, Springer, New York (1995)
30. Du, Q., Faber, V., Gunzburger, M.: Centroidal Voronoi Tessellations: Applications and Algorithms. SIAM Review 41(4), 637–676 (1999)
31. Dunbar, W.B., Murray, R.M.: Distributed Receding Horizon Control with Application to Multi-Vehicle Formation Stabilization. Automatica 42(4), 549–558 (2006)
32. Earl, M.G., D' Andrea, R.: Modeling and Control of a Multi-Agent System using Mixed Integer Linear Programming. In: Proceedings of the IEEE Conference on Decision and Control, December 2002, pp. 107–111 (2002)
33. Elmogy, A.M., Karray, F.O., Khamis, A.M.: Auction-Based Consensus Mechanism for Cooperative Tracking in Multi-Sensor Surveillance Systems. Journal of Advanced Computational Intelligence and Intelligent Informatics 14(1), 13–20 (2010)
34. Erwin, R.S., Albuquerque, P., Jayaweera, S.K., Hussein, I.I.: Dynamic Sensor Tasking for Space Situational Awareness. In: American Control Conference, June/July 2010, pp. 1153–1158 (2010)
35. Fax, J.A., Murray, R.M.: Graph Laplacians and Stabilization of Vehicle Formations. Technical report, California Institute of Technology (2001)

36. Finke, J., Passino, K.M., Sparks, A.: Stable Task Load Balancing Strategies for Cooperative Control of Networked Autonomous Air Vehicles. IEEE Transactions on Control System Technology 14(5), 789–803 (2006)
37. Fiorelli, E., Leonard, N.E., Bhatta, P., Paley, D.A., Bachmayer, R., Fratantoni, D.M.: Multi-AUV Control and Adaptive Sampling in Monterey Bay. In: IEEE Autonomous Underwater Vehicles 2004: Workshop on Multiple AUV Operations (AUV 2004), June 2004, pp. 134–147 (2004)
38. Flanders, H.: Differentiation Under the Integral Sign. The American Mathematical Monthly 80(6), 615–627 (1973)
39. Flint, M., Polycarpou, M., Fernández-Gaucherand, E.: Cooperative Control for Multiple Autonomous UAV's Searching for Targets. In: Proceedings of the 41st IEEE Conference on Decision and Control, December 2002, pp. 2823–2828 (2002)
40. Furukawa, T., Bourgault, F., Lavis, B., Durrant-Whyte, H.F.: Recursive Bayesian Search-and-Tracking using Coordinated UAVs for Lost Targets. In: Proceedings of the 2006 IEEE International Conference on Robotics and Automation, May 2006, pp. 2521–2526 (2006)
41. Furukawa, T., Durrant-Whyte, H.F., Lavis, B.: The Element-based Method - Theory and its Application to Bayesian Search and Tracking. In: IEEE/RSJ International Conference on Intelligent Robots and Systems, October 2007, pp. 2807–2812 (2007)
42. Ganapathy, S., Passino, K.M.: Distributed Agreement Strategies for Cooperative Control: Modeling and Scalability Analysis. In: Butenko, S., Murphey, R., Pardalos, P.M. (eds.) Recent Developments in Cooperative Control and Optimization. Kluwer Academic Publishers (2004)
43. Ganguli, A., Susca, S., Martínez, S., Bullo, F., Cortés, J.: On Collective Motion in Sensor Networks: Sample Problems and Distributed Algorithms. In: IEEE Conference on Decision and Control (December 2005)
44. Gelb, A. (ed.): Applied Optimal Estimation, pp. 107–119. MIT Press, Cambridge (1974)
45. Gerkey, B.P., Matarić, M.J.: A Formal Framework for the Study of Task Allocation in Multi-Robot Systems. International Journal of Robotics Research 23(9), 939–954 (2004)
46. Ghosh, M., Mukhopadhyay, N., Sen, P.K.: Sequential Estimation, 2nd edn. Wiley Series in Probability and Statistics - Applied Probability and Statistics Section Series, vol. 117. Wiley-interscience (1997)
47. Gill, M., Zomaya, A.: Obstacle Avoidance in Multi-Robot Systems: Experiments in Parallel Genetic Algorithms. In: Robotics and Intelligent Systems, vol. 20. World Scientific (1998)
48. Godsil, C., Royle, G.: Algebraic Graph Theory. In: Graduate Texts in Mathematics, vol. 207. Springer, New York (2001)
49. Grocholsky, B.: Information-Theoretic Control of Multiple Sensor Platforms. Ph.D. dissertation, The University of Sydney (2002)
50. Grocholsky, B., Durrant-Whyte, H., Gibbens, P.: An Information-Theoretic Approach to Decentralized Control of Multiple Autonomous Flight Vehicles. In: SPIE Proceedings Series, International Society for Optical Engineering Proceedings Series, vol. 4196, pp. 348–359 (2000)
51. Grocholsky, B., Makarenko, A., Durrant-Whyte, H.: Information-Theoretic Coordinated Control of Multiple Sensor Platforms. In: International Conference on Robotics and Automation, Sepetember 2003, pp. 1521–1526 (2003)

52. Hanselmann, T., Morelande, M., Moran, B., Sarunic, P.: Sensor Scheduling for Multiple Target Tracking and Detection using Passive Measurements. In: 11th International Conference on Information Fusion, pp. 1–8 (2008)

53. Hernandez, M.L., Kirubarajan, T., Bar-Shalom, Y.: Multisensor Resource Deployment Using Posterior Cramér-Rao Bounds. IEEE Transactions on Aerospace and Electronic Systems 40(2), 399–416 (2004)

54. Hero, A.O., Ma, B., Michel, O., Gorman, J.: Alpha-Divergence for Classification, Indexing and Retrieval. Technical report cspl-328, Communications and Signal Processing Laboratory (May 2001)

55. Hert, S., Tiwari, S., Lumelsky, V.: A Terrain-Covering Algorithm for an AUV. Autonomous Robots 3(2/3), 91–119 (1996)

56. Horne, J.K.: Fisheries and Marine Mammal Opportunities in Ocean Observations. In: Heraklion, C. (ed.) Proceedings of Underwater Acoustic Measurements: Technologies & Results (2005)

57. Hussein, I.I.: Motion Planning for Multi-Spacecraft Interferometric Imaging System. Ph.d. dissertation, University of Michigan, Ann Arbor (2005)

58. Hussein, I.I.: A Kalman-Filter Based Control Strategy for Dynamic Coverage Control. In: Proceedings of the American Control Conference, pp. 3271–3276 (2007)

59. Hussein, I.I., Stipanović, D.: Effective Coverage Control using Dynamic Sensor Networks. In: 2006 IEEE Conference on Decision and Control, December 2006, pp. 2747–2752 (2006)

60. Hussein, I.I., Stipanović, D.: Effective Coverage Control for Mobile Sensor Networks with Guaranteed Collision Avoidance. IEEE Transactions on Control Systems Technology, Special Issue on Multi-Vehicle Systems Cooperative Control with Applications 15(4), 642–657 (2007)

61. Hussein, I.I., Stipanović, D.: Effective Coverage Control using Dynamic Sensor Networks with Flocking and Guaranteed Collision Avoidance. In: 2007 American Control Conference, July 2007, pp. 3420–3425 (2007)

62. Hussein, I.I., Stipanovíc, D.M., Wang, Y.: Reliable Coverage Control using Heterogeneous Vehicles. In: 46th IEEE Conference on Decision and Control, December 2007, pp. 6142–6147 (2007)

63. Istepanian, R., Stojanovic, M. (eds.): Underwater Acoustic Digital Signal Processing and Communication Systems, 1st edn. Springer (2002)

64. Jacques, D.R.: Search, Classification and Attack Decisions for Cooperative Wide Area Search Munitions. In: Butenko, S., Murphey, R., Pardalos, P.M. (eds.) Cooperative Control: Models, Applications and Algorithms, pp. 75–93. Kluwer Academic Publishers (2003)

65. Jadbabaie, A., Lin, J., Morse, A.S.: Coordination of Groups of Mobile Autonomous Agents Using Nearest Neighbor Rules. IEEE Transactions on Automatic Control 48(6), 988–1001 (2003)

66. Jin, Y., Liao, Y., Minai, A.A., Polycarpou, M.M.: Balancing Search and Target Response in Cooperative Unmanned Aerial Vehicle (UAV) Teams. IEEE Transactions on Systems, Man, and Cybernetics - Part B: Cybernetics 36(3), 571–587 (2006)

67. Kemp, M., Hobson, B., Meyer, J., Moody, R., Pinnix, H., Schulz, B.: MASA: A Multi – AUV Underwater Search and Data Acquisition System. In: Oceans 2002 MTS/IEEE, October 2002, vol. 1, pp. 311–315 (2002)

68. Khatib, O.: Real-Time Obstacle Avoidance for Manipulators and Mobile Robots. The International Journal of Robotics Research 5(1), 90–98 (1986)

69. Klein, D.J., Morgansen, K.A.: Controlled Collective Motion for Trajectory Tracking. In: Proceedings of the 2006 American Control Conference, June 2006, pp. 5269–5275 (2006)

70. Kreucher, C., Hero III, A.O., Kastella, K.: A Comparison of Task Driven and Information Driven Sensor Management for Target Tracking. In: 44th IEEE Conference on Decision and Control, and the European Control Conference, December 2005, pp. 4004–4009 (2005)
71. Latombe, J.-C.: Robot Motion Planning. In: The Kluwer International Series in Engineering and Computer Science, vol. 124. Kluwer, Boston (1991)
72. Lavis, B., Furukawa, T., Durrant-Whyte, H.F.: Dynamic Space Reconfiguration for Bayesian Search-and-Tracking with Moving Targets. Autonomous Robots 24, 387–399 (2008)
73. Léchevin, N., Rabbath, C.A., Lauzon, M.: A Decision Policy for the Routing and Munitions Management of Multiformations of Unmanned Combat Vehicles in Adversarial Urban Environments. IEEE Transactions on Control Systems Technology 17(3), 505–519 (2009)
74. Leonard, J.J., Durrant-Whyte, H.F.: Simultaneous Map Building and Localization for an Autonomous Mobile Robot. In: IEEE/RSJ International Workshop on Intelligent Robots and Systems IROS 1991, Osaka, Japan, November 1991, pp. 1442–1447 (1991)
75. Leonard, N.E., Paley, D., Lekien, F., Sepulchre, R., Fratantoni, D.M., Davis, R.: Collective Motion, Sensor Networks and Ocean Sampling. In: Proceedings of the IEEE, Special Issue on Networked Control Systems, vol. 95(1), pp. 48–74 (January 2007)
76. Li, W., Cassandras, C.G.: Distributed Cooperative Coverage Control of Sensor Networks. In: Proceedings of the IEEE Conference on Decision and Control, pp. 2542–2547 (2005)
77. Liu, J., Reich, J., Zhao, F.: Collaborative In-Network Processing for Target Tracking. EURASIP Journal of Applied Signal Processing 4, 378–391 (2003)
78. Lloyd, S.: Least Squares Quantization in PCM. IEEE Transactions in Information Theory 28(2), 129–137 (1982)
79. Lumelsky, V., Mukhopadhyay, S., Sun, K.: Dynamic Path Planning in Sensor-Based Terrain Acquisition. IEEE Transactions on Robotics and Automation 6(4), 462–472 (1990)
80. Mahler, R.: Multisensor-Multitarget Sensor Management: A Unified Bayesian Approach. In: Signal Processing, Sensor Fusion, and Target Recognition XII, vol. 5096, pp. 222–233 (2003)
81. Mahler, R.: Objective Functions for Bayeisan Control-Theoretic Sensor Management, I: Multitarget First-Moment Approximation. In: Proceedings of IEEE Aerospace Conference, pp. 1905–1923 (2003)
82. Mak, L.C., Kumon, M., Whitty, M., Katupitiya, J., Furukawa, T.: Design and Development of Micro Aerial Vehicles and their Cooperative Systems for Target Search and Tracking. International Journal of Micro Air Vehicles 1(2), 139–153 (2009)
83. Mataric, M.J.: Using Communication to Reduce Locality in Distributed Multi-Agent Learning. Journal of Experimental and Theoretical Artificial Intelligence 10(3), 357–369 (1998)
84. McLain, T.W., Chandler, P.R., Pachter, M.: A Decomposition Strategy for Optimal Coordination of Unmanned Air Vehicles. In: Proceedings of the American Control Conference, June 2000, pp. 369–373 (2000)
85. Mechitov, K., Sundresh, S., Kwon, Y., Agha, G.: Cooperative Tracking with Binary-Detection Sensor Networks. Technical report, University of Illinois at Urbana-Champaign (September 2003)
86. Meguerdichian, S., Koushanfar, F., Potkonjak, M., Srivastava, M.B.: Coverage Problems in Wireless Ad-hoc Sensor Networks. In: Proceedings of IEEE INFOCOM, pp. 1380–1387 (2001)

87. Mihaylova, L., Lefebvre, T., Bruyninckx, H., Gadeyne, K.: Active Sensing for Robotics-A Survey. In: Proceedings of the 5th International Conference on Numerical Methods and Applications, pp. 316–324 (2002)
88. Miller, J.G.: A New Sensor Allocation Algorithm for the Space Surveillance Network. Military Operations Research 12, 57–70 (2007)
89. Mindell, D., Bingham, B.: New Archaeological Uses of Autonomous Underwater Vehicles. In: MTS/IEEE Conference and Exhibition, OCEANS 2001, vol. 1, pp. 555–558 (2001)
90. Montemerlo, M., Thrun, S., Koller, D., Wegbreit, B.: FastSLAM: A Factored Solution to the Simultaneous Localization and Mapping Problem. In: Proceedings of the AAAI National Conference on Artificial Intelligence, pp. 593–598 (2002)
91. Morris, S., Frew, E.W., Jones, H.: Cooperative Tracking of Moving Targets by Teams of Autonomous Unmanned Air Vehicles. Technical report, University of Colorado at Boulder (July 2005)
92. Mottaghi, R., Payandeh, S.: An Overview of a Probabilistic Tracker for Multiple Cooperative Tracking Agents. In: International Conference on Advanced Robotics, July 2005, pp. 888–894 (2005)
93. Murino, V., Trucco, A.: Three-Dimensional Image Generation and Processing in Underwater Acoustic Vision. Proceedings of the IEEE 88(12), 1903–1946 (2000)
94. Murino, V., Trucco, A., Regazzoni, C.S.: A Probabilistic Approach to the Coupled Reconstruction and Restoration of Underwater Acoustic Images. IEEE Transaction on Pattern Analysis and Machine Intelligence 20(1), 9–22 (1998)
95. Murray, R.M.: Recent Research in Cooperative Control of Multivehicle Systems. Journal of Dynamic Systems, Measurement, and Control 129(5), 571–583 (2007)
96. Nygard, K.E., Chandler, P.R., Pachter, M.: Dynamic Network Flow Optimization Models for Air Vehicle Resource Allocation. In: American Control Conference, June 2001, pp. 1853–1858 (2001)
97. Oh, Y.-J., Watanabe, Y.: Development of Small Robot for Home Floor Cleaning. In: The 41st SICE Annual Conference, August 2002, pp. 3222–3223 (2002)
98. Ohya, A., Kosaka, A., Kak, A.: Vision-Based Navigation by a Mobile Robot with Obstacle Avoidance using Single-Camera Vision and Ultrasonic Sensing. IEEE Transactions on Robotics and Automation 14(6), 969–978 (1998)
99. Okabe, A., Suzuki, A.: Locational Optimization Problems Solved through Voronoi Diagrams. European Journal of Operational Research 98(3), 445–456 (1997)
100. Olfati-Saber, R., Murray, R.M.: Distributed Cooperative Control of Multiple Vehicle Formations. In: Proceedings of the 15th IFAC World Congress (July 2002)
101. Olfati-Saber, R., Murray, R.M.: Consensus Problems in Networks of Agents With Switching Topology and Time-Delays. IEEE Transactions on Automatic Control 49(9), 1520–1533 (2004)
102. Ollero, A., Martínez de Dios, J.R., Merino, L.: Unmanned Aerial Vehicles as Tools for Forest-Fire Fighting. In: V International Conference on Forest Fire Research (2006)
103. Page, E.S.: Continuous Inspection Scheme. Biometrika 41(1/2), 100–115 (1954)
104. Parker, L.: Cooperative Robotics for Multi-Target Observation. Intelligent Automation and Soft Computing 5(1), 5–19 (1999)
105. Poor, H.V.: An Introduction to Signal Detection and Estimation, 2nd edn. Springer (1994)
106. Poor, H.V., Hadjiliadis, O.: Quickest Detection, 1st edn. Cambridge University Press (December 2008)
107. Ren, W., Beard, R.W., Kingston, D.B.: Multi-Agent Kalman Consensus with Relative Uncertainty. In: Proceedings of the American Control Conference, pp. 1865–1870 (2005)

108. Rényi, A.: On Measures of Information and Entropy. In: Proceedings of the 4th Berkeley Symposium on Mathematics, Statistics and Probability, pp. 547–561 (1961)
109. Roberts, S.W.: A Comparison of Some Control Chart Procedures. Technometrics 8(3), 411–430 (1966)
110. Ross, S., Pineau, J., Paquet, S., Chaib-draa, B.: Online Planning Algorithms for POMDPs. Journal of Artificial Intelligence Research 32(2), 663–704 (2008)
111. Rumsfeld, D.H.: Report of the Commission to Assess United States National Security Space Management and Organization. Technical report, United States Congress (January 2001)
112. Schonhoff, T.A., Giordano, A.A.: Detection and Estimation Theory and Its Applications. Prentice Hall (2006)
113. Shiryaev, A.N.: Optimal Stopping Rules. Springer (November 2007)
114. Sinha, A., Kirubarajan, T., Bar-Shalom, Y.: Autonomous Ground Target Tracking by Multiple Cooperative UAVs. In: IEEE Aerospace Conference, pp. 1–9 (2005)
115. Smith, T.: Probabilistic Planning for Robotic Exploration. PhD thesis, Carnegie Mellon University (July 2007)
116. Spires, S., Goldsmith, S.: Exhaustive Geographic Search with Mobile Robots Along Space-Filling Curves. In: 1st International Workshop on Collective Robotics, pp. 1–12 (1998)
117. Spletzer, J.R., Taylor, C.J.: Dynamic Sensor Planning and Control for Optimally Tracking Targets. The International Journal of Robotics Research (1), 7–20 (2003)
118. Sujit, P.B., Beard, R.: Distributed Sequential Auctions for Multiple UAV Task Allocation. In: Proceedings of 2007 American Control Conference, July 2007, pp. 3955–3960 (2007)
119. Sujit, P.B., Ghose, D.: Multiple UAV Search using Agent Based Negotiation Scheme. In: American Control Conference, June 2005, pp. 2995–3000 (2005)
120. Svennebring, J., Koenig, S.: Building Terrain-Covering Ant Robots: A Feasibility Study. Autonomous Robots 16(3), 313–332 (2004)
121. Tanner, H.G., Jadbabaie, A., Pappas, G.J.: Stable Flocking of Mobile agents, Part I: Fixed Topology. In: 42nd IEEE Conference on Decision and Control, December 2003, pp. 2010–2015 (2003)
122. Tanner, H.G., Jadbabaie, A., Pappas, G.J.: Stable Flocking of Mobile agents, Part II: Dynamic Topology. In: 42nd IEEE Conference on Decision and Control, December 2003, pp. 2016–2021 (2003)
123. Tanner, H.G., Jadbabaie, A., Pappasa, G.J.: Flocking in Fixed and Switching Networks. Transactions on Automatic Control 52(5), 863–868 (2007)
124. Thrun, S., Burgard, W., Fox, D.: Probabilistic Robotics. In: Intelligent Robotics and Autonomous Agents. The MIT Press (September 2005)
125. Tisdale, J., Ryan, A., Kim, Z., Törnqvist, D., Hedrick, J.K.: A Multiple UAV System for Vision-Based Search and Localization. In: American Control Conference, June 2008, pp. 1985–1990 (2008)
126. Van Trees, H.L.: Detection, Estimation, and Modulation Theory, vol. I. John Wiley & Sons (2003)
127. Vidal, R., Shakernia, O., Kim, H.J., Shim, D.H., Sastry, S.: Probabilistic Pursuit-Evasion Games: Theory, Implementation and Experimental Evaluation. IEEE Transactions on Robotics and Automation 18(5), 662–669 (2002)
128. Viswanathan, R., Varshney, P.K.: Distributed Detection with Multiple Sensors: Part 1 - Fundamentals. Proc. IEEE 85(1), 54–63 (1997)
129. von Alt, C.: REMUS 100 Transportable Mine Countermeasure Package. In: Oceans 2003 Proceedings, San Diego, CA, September 2003, vol. 4, pp. 1925–1930 (2003)

130. Wagner, I., Lindenbaum, M., Bruckstein, A.: Distributed Covering by Ant-Robots using Evaporating Traces. IEEE Transactions on Robotics and Automation 15(5), 918–933 (1999)
131. Wald, A.: Sequential Tests of Statistical Hypotheses. The Annals of Mathematical Statistics 16(2), 117–186 (1945)
132. Wald, A.: Sequential Analysis. Dover Publications (2004)
133. Wald, A., Wolfowitz, J.: Bayes Solutions of Sequential Decision Problems. The Annals of Mathematical Statistics 21(1), 82–99 (1950)
134. Wang, B.: Coverage Control in Sensor Networks. Springer (2010)
135. Wang, Y., Hussein, I.I.: Cooperative Vision-based Multi-vehicle Dynamic Coverage Control for Underwater Applications. In: Proceedings of the IEEE Multiconference on Systems and Control, pp. 82–87 (2007) (invited paper)
136. Wang, Y., Hussein, I.I.: Underwater Acoustic Imaging using Autonomous Vehicles. In: 2008 IFAC Workshop on Navigation, Guidance and Control of Underwater Vehilces (April 2008)
137. Wang, Y., Hussein, I.I.: Awareness Coverage Control Over Large Scale Domains with Intermittent Communications. IEEE Transactions on Automatic Control 55(8), 1850–1859 (2010)
138. Wang, Y., Hussein, I.I.: Bayesian-Based Decision-Making for Object Search and Classification. IEEE Transactions on Control Systems Technology (2010) (in press)
139. Wang, Y., Hussein, I.I., Brown III, D.R., Erwin, R.S.: Cost-Aware Sequential Bayesian Tasking and Decision-Making for Search and Classification. In: American Control Conference, pp. 6423–6428 (2010)
140. Wang, Y., Hussein, I.I., Erwin, R.S.: Awareness-Based Decision Making for Search and Tracking. In: American Control Conference, pp. 3169–3175 (2008) (invited paper)
141. Wang, Y., Hussein, I.I., Erwin, R.S.: Bayesian Detection and Classification for Space-Augmented Space Situational Awareness under Intermittent Communications. In: Military Communications Conference (MILCOM), October 2010, pp. 960–965 (2010)
142. Whalen, A.D.: Detection of Signals in Noise. Academic Press, New York (1971)
143. Winfield, A.: Distributed Sensing and Data Collection via Broken Ad Hoc Wireless Connected Networks of Mobile Robots. In: Proceedings of 5th International Symposium on Distributed Autonomous Robotic Systems, pp. 273–282 (2000)
144. Wintenby, J., Krishnamurthy, V.: Hierarchical Resource Management in Adaptive Airborne Surveillance Radars. IEEE Transactions on Aerospace and Electronic Systems 42(2), 401–420 (2006)
145. Wong, S., MacDonald, B.: A Topological Coverage Algorithm for Mobile Robots. In: IEEE/RSJ International Conference on Intelligent Robots and Systems, October 2003, pp. 1685–1690 (2003)
146. Yang, S., Luo, C.: A Neural Network Approach to Complete Coverage Path Planning. IEEE Transactions on Systems, Man, and Cybernetics - Part B: Cybernetics 34(1), 718–724 (2004)
147. Yang, Y., Polycarpou, M.M., Minai, A.A.: Multi-UAV Cooperative Search using an Opportunistic Learning Method. Journal of Dynamic Systems, Measurement, and Control 129(5), 716–728 (2007)
148. Yu, X., Azimi-Sadjadi, M.R.: Neural Network Directed Bayes Decision Rule for Moving Target Classification. IEEE Transactions on Aerospace and Electronic Systems 36(1), 176–188 (2000)

Index